Roberto Heredia Ortiz

Modélisation toxicocinétique du benzo(a)pyrène

AF191670

Roberto Heredia Ortiz

Modélisation toxicocinétique du benzo(a)pyrène

L'interprétation des données de surveillance biologique de l'exposition chez les travailleurs

Presses Académiques Francophones

Impressum / Mentions légales
Bibliografische Information der Deutschen Nationalbibliothek: Die Deutsche Nationalbibliothek verzeichnet diese Publikation in der Deutschen Nationalbibliografie; detaillierte bibliografische Daten sind im Internet über http://dnb.d-nb.de abrufbar.
Alle in diesem Buch genannten Marken und Produktnamen unterliegen warenzeichen-, marken- oder patentrechtlichem Schutz bzw. sind Warenzeichen oder eingetragene Warenzeichen der jeweiligen Inhaber. Die Wiedergabe von Marken, Produktnamen, Gebrauchsnamen, Handelsnamen, Warenbezeichnungen u.s.w. in diesem Werk berechtigt auch ohne besondere Kennzeichnung nicht zu der Annahme, dass solche Namen im Sinne der Warenzeichen- und Markenschutzgesetzgebung als frei zu betrachten wären und daher von jedermann benutzt werden dürften.

Information bibliographique publiée par la Deutsche Nationalbibliothek: La Deutsche Nationalbibliothek inscrit cette publication à la Deutsche Nationalbibliografie; des données bibliographiques détaillées sont disponibles sur internet à l'adresse http://dnb.d-nb.de.
Toutes marques et noms de produits mentionnés dans ce livre demeurent sous la protection des marques, des marques déposées et des brevets, et sont des marques ou des marques déposées de leurs détenteurs respectifs. L'utilisation des marques, noms de produits, noms communs, noms commerciaux, descriptions de produits, etc, même sans qu'ils soient mentionnés de façon particulière dans ce livre ne signifie en aucune façon que ces noms peuvent être utilisés sans restriction à l'égard de la législation pour la protection des marques et des marques déposées et pourraient donc être utilisés par quiconque.

Coverbild / Photo de couverture: www.ingimage.com

Verlag / Editeur:
Presses Académiques Francophones
ist ein Imprint der / est une marque déposée de
OmniScriptum GmbH & Co. KG
Heinrich-Böcking-Str. 6-8, 66121 Saarbrücken, Deutschland / Allemagne
Email: info@presses-academiques.com

Herstellung: siehe letzte Seite /
Impression: voir la dernière page
ISBN: 978-3-8416-3266-1

Zugl. / Agréé par: Montréal, Université de Montréal, 2014

Livre intitulé :

Modélisation toxicocinétique du benzo[a]pyrène

L'interprétation des données de surveillance biologique de l'exposition chez les

travailleurs

Présenté par :

Roberto Heredia Ortiz

Résumé

De nombreux travailleurs sont exposés aux hydrocarbures aromatiques polycycliques (HAP). Le benzo(a)pyrène (BaP) fait partie de ce groupe de polluants. Cette substance a été classée cancérogène reconnu chez l'humain. Pour évaluer l'exposition aux HAP cancérogènes, plusieurs chercheurs ont proposé d'utiliser la mesure du 3-hydroxybenzo(a)pyrène (3-OHBaP) dans l'urine des travailleurs exposés. Dans le cadre du présent projet, deux approches de modélisation ont été développées et appliquées pour permettre une meilleure compréhension de la toxicocinétique du BaP et son biomarqueur d'intérêt actuel, le 3-OHBaP, et pour aider à interpréter les résultats de surveillance biologique.

Un modèle toxicocinétique à plusieurs compartiments a été développé sur la base des données préalablement obtenues sur le rat par notre groupe. Selon le modèle, le BaP injecté par voie intraveineuse est rapidement distribué du sang vers les tissus ($t_{\frac{1}{2}} \approx 4$ h), avec une affinité particulière pour les poumons et les composantes lipidiques des tissus. Le BaP est ensuite distribué vers la peau et le foie. Au foie, le BaP est promptement métabolisé et le 3-OHBaP est formé avec une demi-vie de ≈ 3 h. Le métabolisme pulmonaire du BaP a également été pris en compte, mais sa contribution à la cinétique globale du BaP a été jugée négligeable. Une fois formé, le 3-OHBaP est distribué vers les différents organes presque aussi rapidement que la molécule mère ($t_{\frac{1}{2}} \approx 2$ h). Le profil temporel du 3-OHBaP dans le rein montre une accumulation transitoire en raison de la différence observée entre le taux d'entrée ($t_{\frac{1}{2}} = 28$ min) et le taux de sortie ($t_{\frac{1}{2}} = 4,5$ h). La clairance totale de 3-OHBaP du corps est principalement gouvernée par le taux de transfert de la bile vers le tractus gastro-intestinal ($t_{\frac{1}{2}} \approx 4$ h). Le modèle toxicocinétique à plusieurs compartiments a réussi à simuler un ensemble indépendant de profils urinaires publiés sur le 3-OHBaP. Ce modèle toxicocinétique à compartiments s'est avéré utile pour la determination des facteurs biologiques déterminants de la cinétique du BaP et du 3-OHBaP.

Par la suite, un modèle pharmacocinétique à base physiologique (PCBP) reproduisant le devenir du BaP et du 3-OHBaP chez le rat a été construit. Les organes (ou tissus) représentés comme des compartiments ont été choisis en fonction de données expérimentales obtenues *in vivo* chez le rat. Les coefficients de partition, les coefficients de perméabilité, les taux de métabolisation, les paramètres d'excrétion, les fractions absorbées et les taux d'absorption pour différentes voies d'exposition ont été obtenus directement à partir des profils sanguins, tissulaires, urinaires et fécaux du BaP et du 3-OHBaP. Les valeurs de ces derniers paramètres ont été calculées par des procédures Monte-Carlo. Des analyses de sensibilité ont ensuite été réalisées pour s'assurer de la stabilité du modèle et pour établir les paramètres les plus sensibles de la cinétique globale. Cette modélisation a permis d'identifier les facteurs déterminants de la cinétique: 1) la sensibilité élevée des paramètres de la métabolisation hépatique du BaP et du 3-OHBaP ainsi que du taux d'élimination; 2) la forte distribution du BaP dans les poumons par rapport à d'autres tissus; 3) la distribution considérable du BaP dans les tissus adipeux et le foie; 4) la forte distribution du 3-OHBaP dans les reins; 5) le transfert limité du BaP par la diffusion tissulaire dans les poumons; 6) le transfert limité du 3-OHBaP par la diffusion tissulaire dans les poumons, les tissus adipeux et les reins; 7) la recirculation entéro-hépatique significative du 3-OHBaP. Suite à des analyses de qualité des ajustements des équations du modèle aux données observées, les probabilités que les simulations reproduisent les données expérimentales par pur hasard se sont avérées toujours inférieures à 10% pour les quatre voies d'exposition : intraveineuse, orale, cutanée et respiratoire.

Nous avons extrapolé les modèles cinétiques du rat à l'humain afin de se doter d'un outil permettant de reconstituer les doses absorbées chez des travailleurs exposés dans diverses industries à partir de mesures de l'évolution temporelle du 3-OHBaP dans leur urine. Les résultats de ces modélisations ont ensuite été comparés à ceux de simulations obtenues avec un modèle toxicocinétique à compartiment unique pour vérifier l'utilité comparative d'un modèle simple et complexe. Les deux types de modèle ont ainsi été construits à partir de profils sanguins, tissulaires, urinaires et fécaux du BaP et du 3-OHBaP sur des rats exposés. Ces données ont été obtenues *in vivo* par voie intraveineuse, cutanée, respiratoire et orale. Ensuite, les modèles ont été extrapolés à l'humain en tenant compte des déterminants biologiques essentiels des différences cinétiques entre le rat et l'humain. Les résultats ont montré que l'inhalation n'était pas la principale voie d'exposition pour plusieurs travailleurs étudiés. Les valeurs de concentrations de BaP dans l'air utilisées afin de simuler les profils d'excrétion urinaire chez les travailleurs étaient différentes des valeurs de concentrations de BaP mesurées dans l'air. Une exposition au BaP par voie cutanée semblait mieux prédire les profils temporels observés. Finalement, les deux types de modélisation se sont avérés utiles pour reproduire et pour interpréter les données disponibles chez des travailleurs.

Mots-clés : Modélisation cinétique, surveillance biologique, dosimétrie inverse, hydrocarbures aromatiques polycycliques, benzo(a)pyrène, 3-hydroxybenzo(a)pyrène.

Abstract

Many workers are exposed to polycyclic aromatic hydrocarbons (PAHs). Benzo(a) pyrene (BaP) is part of this group of pollutants. This substance has been classified as a known carcinogen in humans. To assess exposure to carcinogenic PAHs, several researchers have proposed using the measurement of 3-hydroxybenzo(a)pyrene (3-OHBaP) in the urine of exposed workers. In this project, two modeling approaches were developed and applied to enable a better understanding of the toxicokinetics of BaP and its biomarker of current interest, 3-OHBaP, to help interpret the results of biological monitoring.

A multi-compartment toxicokinetic model was developed based on the data previously obtained in rats by our group of research. According to the model, BaP injected intravenously is rapidly distributed from blood to tissues ($t_{1/2} \approx 4$ h), with a particular affinity for lungs and lipid components of tissues. Subsequently, BaP is distributed to the liver and the skin. Once in the liver, BaP is promptly metabolized and 3-OHBaP is formed with a half-life of about 3 h. Pulmonary biotransformation of BaP was also taken into account, but its contribution to the overall kinetics of BaP was considered negligible. Once formed, 3-OHBaP is distributed to various organs almost as fast as the parent compound ($t_{1/2} \approx 2$ h). An accumulation of 3-OHBaP profile is present in the kidneys because of the difference between the uptake rate ($t_{1/2} = 28$ min) and the ouput rate ($t_{1/2} = 4.5$ h). Total clearance of 3-OHBaP from the blood stream is primarily governed by the rate of transfer of the bile to the gastrointestinal tract ($t_{1/2} \approx 4$ h). The multi-compartment toxicokinetic model was able to simulate an independent set of published 3-OHBaP urinary profiles. This toxicokinetic compartmental model has proved useful for the determination of the main biological features of the kinetics of BaP and 3-OHBaP.

Thereafter, a physiological pharmacokinetic model (PBPK) reproducing the fate of BaP and 3-OHBaP rats was built. Organs (or tissues) represented as compartments were selected based on experimental data obtained *in vivo* in rats. Partition coefficients, coefficients of permeability, biotransformation rates, excretion parameters, and absorption fraction for different exposure routes were obtained directly from the profiles of BaP and 3-OHBaP in blood, various tissues and excreta. The values of these parameters were calculated by Monte Carlo procedures. Sensitivity analyses were then performed to ensure the stability of the model and to determine the most sensitive parameters. This modeling has identified the following features: 1) a high sensitivity of hepatic metabolism and elimination rates of BaP and 3-OHBaP; 2) a large distribution of BaP in the lungs compared to other tissues; 3) a considerable distribution of BaP in adipose tissues and liver; 4) a significant distribution of 3-OHBaP in the kidneys; 5) a diffusion-limited transfer of BaP in the lungs, 6) a diffusion-limited transfer of 3-OHBaP in lungs, adipose tissues and kidneys; and 7) a significant entero-hepatic recycling of 3-OHBaP. Following a series of analysis of goodness of fit, the probabilites that the model simulations reproduced the experimental data due to pure chance were always below 10%, for the four routes of exposure: intravenous, oral, dermal and respiratory.

Subsequently, we have extrapolated the kinetic models from rats to humans in order reproduce the temporal evolution of 3-OHBaP biomarker of exposure in the urine of workers occupationally expose. Results of these models were then compared to simulations obtained with a single compartment toxicokinetic model to verify the comparative usefulness of simple and complex model. Both types of models have been constructed from blood, tissue, urinary and faecal profiles of BaP and 3-OHBaP in rats. These data were obtained *in vivo* by intravenous, subcutaneous, oral and respiratory exposure. The models were extrapolated to humans taking into account the essential biological determinants of kinetic differences between rats and humans. Results showed that inhalation was not the primary route of exposure for many workers studied. The values of air concentrations of BaP used to simulate the urinary excretion profiles were different from those measured in the air. Dermal exposure to BaP seemed to better predict the temporal patterns observed. Finally, the two types of modeling have been proved useful to reproduce and to interpret experimental data obtained in workers.

Keywords : Kinetic modeling, biological monitoring, reverse dosimetry, polycyclic aromatic hydrocarbons, benzo(a)pyrene, 3-hydroxybenzo(a)pyrene.

Table des matières

xv

Liste des tableaux

Liste des figures

Liste des sigles et abréviations

4,5-diolBaP :	4,5-dihydrodiol-benzo(a)pyrène
7,8-diolBaP :	7,8-dihydrodiol-benzo(a)pyrène
9,10-diolBaP :	9,10-dihydrodiol-benzo(a)pyrène
1,6-dione-BaP :	1,6-dione-benzo(a)pyrène
3,6-dione-BaP :	3,6-dione-benzo(a)pyrène
7,8-dione-BaP :	7,8-dione-benzo(a)pyrène
1-OHP :	1-hydroxypyrène
3-OHBaP :	3-hydroxybenzo(a)pyrène
6-OHBaP :	6-hydroxybenzo(a)pyrène
7-OHBaP :	7-hydroxybenzo(a)pyrène
9-OHBaP :	9-hydroxybenzo(a)pyrène
9-OH-4,5-diolBaP :	9-hydroxy -4,5-dihydrodiol-benzo(a)pyrène
7,8,9,10-tetraolBaP :	7,8,9,1-tetraol-benzo(a)pyrène
A :	Sang artériel
ADME :	Absorption, distribution, métabolisme et excrétion
ADN	Acide désoxyribonucléique
ANSES :	Agence nationale de sécurité sanitaire de l'alimentation, de l'environnement et du travail
ATSDR :	Agence de substances toxiques et registre des maladies des Etats-Unis
AUC :	Aire sous la courbe

B(t) :	Quantité de xénobiotique dans le corps entier ou dans le sang en fonction du temps
BaP :	Benzo(a)pyrène
BaP-1,3-diOH	Benzo(a)pyrène-1,3-diOH
BaP-7,8-diol-9,10-époxyde :	Benzo(a)pyrène-7,8-diol-9,10-époxyde
BaP-2,3-époxyde :	Benzo(a)pyrène-2,3-époxyde
BaP-4,5-époxyde :	Benzo(a)pyrène-4,5-époxyde
BaP-7,8-époxyde :	Benzo(a)pyrène-7,8-époxyde
BaP-9,10-époxyde :	Benzo(a)pyrène-9,10-époxyde
BaP-9-OH-4,5-époxyde:	Benzo(a)pyrène-9-OH-4,5-époxyde
BaP-1,2-quinone :	Benzo(a)pyrène-1,2-quinone
BaP-3,6-quinone :	Benzo(a)pyrène-3,6-quinone
BaP-7,8-quinone :	Benzo(a)pyrène-7,8-quinone
BaP-6,12-quinone :	Benzo(a)pyrène-6,12-quinone
B :	Bile
CIRC :	Centre international de recherche sur le cancer
Cl :	Clairance
CU :	Cutané
CYP 450 :	Cytochrome(s) P450
CYP1A1 :	Cytochrome P450 1A1
CYP1A2 :	Cytochrome P450 1A2
CYP1B1 :	Cytochrome P450 1B1
CYP3A4 :	Cytochrome P450 3A4

xxii

DMSENO :	Dose maximale sans effet nocif observable
DMENO :	Dose minimale pour un effet nocif observable
D(t) :	Dose administrée en fonction du temps
E :	Échange
EC :	Commission européenne
EPA :	Agence de la protection de l'environnement des Etats-Unis
F :	Foie
FE :	Fèces
F(t) :	Quantité de xénobiotique éliminée par les fèces en fonction du temps
GI(t) :	Quantité de xénobiotique dans le tractus gastro-intestinal en fonction du temps
GSTM :	Glutathion S-transférase
HAP :	Hydrocarbures aromatiques polycycliques
IN :	Inhalation
INRP :	Inventaire national des rejets de polluants
IP :	Intra-péritonéale
IV :	Intraveineuse
k, K_{elim} :	Taux d'élimination
K_{PA} :	Produit de perméabilité
K_m :	Constante de Michaelis-Menten
LD :	Limité par la diffusion
M :	Métabolites

xxiii

NIEHS :	Institut national des sciences de la santé environnementale
NIOSH :	Institut national pour la santé et la sécurité au travail
O :	Oral
OIT :	Organisation internationale du travail
O(t) :	Quantité de xénobiotique éliminée par d'autres voies en fonction du temps
p :	Valeur de signification
P :	Coefficient de partition
PCBP :	Pharmacocinétique à base physiologique
PE :	Peau
POU :	Poumons
Q :	Flux sanguin
SCF :	Comité scientifique sur l'alimentation
$t_{1/2}$:	Demi-vie
TPP :	Tissus pauvrement perfusés
TRP ·	Tissus richement perfusés
UGT :	Uridine diphosphate glucuronyltransférase
U(t) :	Quantité de xénobiotique éliminée par l'urine en fonction du temps
V :	Sang veineux ou volume
V_d :	Volume de distribution apparent
V_{max} :	Vitesse maximale
VRB :	Valeur de référence biologique

À M.L.

Présentation du livre

1 Introduction

Les études actuelles d'évaluation des risques toxicologiques tentent d'établir des outils fiables permettant de reconstituer les doses absorbées de polluants par les individus à partir des mesures de biomarqueurs d'exposition retrouvés dans des échantillons biologiques accessibles (NRC, 2006; Nielsen *et al.*, 2008; Que Hee, 1993). Afin d'évaluer les risques associés à l'exposition à diverses substances toxiques, une bonne connaissance de leur cinétique *in vivo* est d'une importance primordiale (Bouchard 1998 ; Spengler *et al.*, 2000; Tardiff et Goldstein, 1991). Ainsi, des modèles toxicocinétiques ont été développés afin d'établir la relation entre les doses d'exposition de contaminants et les niveaux biologiques de métabolites (Carrier et al. 2001 ; Bouchard et al. 2005 ; Leeuwen et Vermeire, 2007; OIT, 1986). Ces modèles mathématiques permettent de reproduire les profils temporels de biomarqueurs d'exposition quelle que soit la voie d'administration, ce qui les rend grandement utiles pour évaluer le risque dans plusieurs milieux et contextes (NRC, 1987).

Les hydrocarbures aromatiques polycycliques (HAP) font partie des substances chimiques qui présentent une forte toxicité pour l'être humain. Les HAP résultent généralement de processus de combustion et sont présents naturellement dans le pétrole, le charbon et les dépôts de goudron (Haines et Hendrickson, 2009). Le benzo(a)pyrène (BaP) est un HAP parmi les plus communs, connu pour ses effets cancérogènes chez plusieurs espèces (Bostrom *et al.*, 2002; Harvey, 1991; Luch, 2005). Le Centre international de recherche sur le cancer l'a classé parmi les agents cancérogènes reconnus chez humain (correspondant au groupe 1) à partir de l'édition 2010 de sa monographie (Baan *et al.*, 2009; IARC, 2010). Le BaP est rapidement métabolisé par divers organes. Des expériences *in vivo* sur des rats suggèrent que le foie et le poumon sont les principaux organes de biotransformation du BaP (Harrigan *et al.*, 2006). Suite à la biotransformation du BaP, une cascade d'une vingtaine de sous-produits est constituée (Moore et Cohen, 1978). La biotransformation du BaP passe par quatre voies de métabolisme principales. La voie générant des métabolites phénoliques résulte de l'hydrolyse du BaP. Le 3-hydroxybenzo(a)pyrène (3-OHBaP) est l'un de ces métabolites qui est majoritairement éliminé par les fèces (Marie *et al.*, 2010; Ramesh *et al.*, 2001b). Récemment, le 3-OHBaP a été étudié comme indicateur de l'exposition aux HAP cancérogènes dans les milieux de travail à risque et dans la population générale (Ariese *et al.*, 1994; Bouchard *et al.*, 1994; Leroyer *et al.*, 2010).

Dans le présent livre, nous avons émis l'hypothèse que la modélisation toxicocinétique à compartiments et celle à base physiologique sont toutes les deux aptes à relier l'excrétion urinaire du 3-OHBaP aux doses de BaP absorbées chez l'humain. Au chapitre 2, l'état actuel des connaissances a été documenté. Ce chapitre recueille les éléments essentiels pour comprendre la problématique de cette recherche. Ainsi, ce chapitre 2 introduit le BaP, les principales sources d'exposition, la toxicologie associée au BaP, la toxicocinétique correspondante ainsi que les normes et les valeurs de référence. Par la suite, la surveillance biologique de l'exposition au BaP est abordée. Les concepts nécessaires à la modélisation toxicocinétique sont ensuite décrits. Au chapitre 3, la problématique et les objectifs de recherche sont présentés. Aux chapitres 4, 5, et 6, les résultats de notre étude sont présentés sous forme de trois articles de recherche publiés ou acceptés pour publication. Finalement, au chapitre 7, une discussion générale des résultats et une brève conclusion sont présentées.

2 État des connaissances

2.1 Le benzo(a)pyrène

2.1.1 La famille des hydrocarbures aromatiques polycycliques

Les hydrocarbures aromatiques polycycliques font partie d'un groupe de produits lipophiles composés principalement de cycles aromatiques fusionnés (c'est-à-dire, sans la présence d'un seul hétéroatome dans la molécule, Harvey (1997)). Les HAP sont généralement formés lors de la combustion incomplète du pétrole, du gaz, du goudron, du bois ou d'autres substances organiques (Haines et Hendrickson, 2009). Ils se retrouvent naturellement dans des mélanges complexes et sont des sous-produits de la combustion de carburants. Les HAP se présentent généralement sous forme solide incolore, blanc, jaune pâle ou bleu pâle ayant une faible odeur agréable, tel que décrit dans le profil toxicologique des produits du Département américain de la santé et des services sociaux (ATSDR, 1995b). Ils peuvent être trouvés dans l'air, l'eau, le sol, les aliments et être largement distribués dans l'environnement (Allamandola, 2011). Des HAP sont présents dans certains médicaments, colorants, matières plastiques, pesticides, asphalte, *etc.* Il a été constaté que, pour la grande majorité de la population américaine, les principales sources d'exposition aux HAP sont l'inhalation de la fumée de tabac ou de bois, de l'air ambiant ainsi que la consommation d'aliments (ATSDR, 1995a). En raison de leurs éventuels effets cancérogènes, mutagènes et tératogènes, observés chez les animaux (Harvey, 1991; Luch, 2005), plusieurs HAP ont soulevé l'intérêt des organismes de réglementation en matière de santé publique. Le benzo(a)pyrène (BaP) a été identifié comme prioritaire étant donné ses effets toxiques sur l'humain et le fait qu'il soit représentatif des HAP cancérogènes (IARC, 2010).

2.1.2 Historique

En 1918, des chercheurs japonais ont constaté l'apparition de cancers chez des lapins exposés au goudron de houille (Yamagiwa et Ichikawa, 1977). Cette étude a pu confirmer les données épidémiologiques observées par Pott sur le cancer du scrotum chez les ramoneurs plusieurs décennies auparavant (Rescigno, 2010). En 1933, des chercheurs britanniques ont identifié les principaux HAP dont le BaP. Dans leurs études, ils ont démontré la cancérogénicité du BaP lors de l'exposition cutanée de souris (Yang et Silverman, 1988). En 1947, les docteurs Miller ont démontré que plusieurs substances cancérogènes devaient être métabolisées par l'organisme pour produire des métabolites électrophiles pouvant former des adduits covalents (selon Loeb et Harris, 2008). En 1956, Conney, en collaboration avec les docteurs Miller, a identifié les enzymes microsomales comme étant les principales responsables de l'activation d'autres HAP cancérogènes chez le rat (Conney *et al.*, 1956). Sur la base de leur présence dans l'environnement et de potentiel toxique prouvé, seize HAP ont été sélectionnés par l'Agence de protection environnementale des États-Unis dans leur liste de polluants prioritaires en 2014 : le naphtalène, l'acénaphtylène, l'acénaphtène, le fluorène, le phénanthrène, l'anthracène, le fluoranthène, le pyrène, le benzo(a)anthracène, le chrysène, le benzo(b)fluoranthène, le benzo(k)fluoranthène, le benzo(a)pyrène, l'indéno (1,2,3-cd)pyrène, le dibenzo (ah) anthracène et le benzo(ghi)pérylène (Coelho *et al.*, 2008; EPA, 2014; Santos et Galceran, 2002). Le benzo(a)anthracène, le benzo(a)pyrène, le benzo(b)fluoranthène, le benzo(k)fluoranthène, le chrysène, le dibenzo(a,h)anthracène, et l'indéno(1,2,3-c,d)pyrène sont considérés comme cancérigènes probables ou reconnus pour l'humain (EPA, 2014). En 1974, Sims et ses collaborateurs ont documenté la formation du benzo(a)pyrène 7,8-diol-9,10-epoxide (BaP-7,8-diol-9,10-époxyde) et sa liaison à l'ADN (Sims *et al.*, 1974). Ceci a permis d'établir que l'ADN est une cible cellulaire des HAP cancérogènes et ainsi établir la relation entre mutations et cancer. Par la suite, un ensemble d'études a été réalisé pour mieux comprendre cette relation. Dans son rapport de 1983, le Centre international sur la recherche

4

du cancer (CIRC) avait déjà classé le BaP dans le groupe 2A des agents possiblement cancérogènes pour l'être humain (IARC, 1983). La structure des principaux adduits à l'ADN du BaP a été définie en 1997 (Carrell *et al.*, 1997). En 2010, le BaP a été classé parmi les agents cancérogènes reconnus chez l'humain par le CIRC (IARC, 2010).

2.1.3 Caractéristiques physico-chimiques

Le nom officiel du BaP selon l'Union internationale de chimie pure et appliquée est benzo(a)pyrène et benzo[a]pyrène. Son numéro de CAS est le 50-32-8. Ses principales caractéristiques physico-chimiques sont résumées au Tableau 2-1 (IARC, 1983; IARC, 2010).

Tableau 2-1 : Principales caractéristiques physico-chimiques du BaP [a].

Propriété	
Formule moléculaire	$C_{20}H_{12}$
Structure chimique	
Numéro de registre	50-32-8
Noms et autres abréviations communes	B(a)P
	BAP
	3,4-bp
	Bp
	Benzo[def]chrysène
	3,4-benzopyrène
	3,4-benzpyrène

5

Propriété	
	3,4-benz(a)pyrène benz(a)pyrène
Poids moléculaire (g/mol)	252,3
État physique	Solide, jaune pâle
Odeur	Légère odeur aromatique
Point de fusion (°C)	178,1
Point d'ébullition (°C)	310-312
Densité (g/cm³)	1,351
Solubilité dans l'eau (% de la masse)	Insoluble, 3×10^{-7} à 25 °C
Solubilité dans d'autres milieux	Très soluble dans le chloroforme, soluble dans le benzène, le toluène et le xylène
Constante de la loi de Henry (kPa m³/mol)	$4,65 \times 10^{-5}$ à 25 °C
Point d'éclair	Ininflammable

[a] Données obtenues à partir de l'IARC (1983).

2.1.4 Les sources d'exposition

2.1.4.1 Les sources naturelles

Les HAP sont produits naturellement dans des mélanges complexes. Le BaP est produit lors de la combustion ou de la pyrolyse des matières organiques. Les feux de forêts et les éruptions volcaniques constituent des sources naturelles importantes de HAP. Le type de matière, le temps de production, la disponibilité de l'oxygène, la température, la pression et d'autres conditions réactives, déterminent la quantité et la variété des HAP produits. Donc, la proportion du BaP par rapport aux autres HAP dépend de toutes ces propriétés de formation (Santé Canada, 1994). Les HAP sont également retrouvés de façon naturelle dans les sédiments souterrains et originent de la formation des roches (diagenèse). Les HAP peuvent être formés lors de réactions de biosynthèse (il a été documenté que le BaP peut être synthétisé par plusieurs bactéries et algues (Verschueren, 2001).

2.1.4.2 Les sources artificielles

Actuellement, il n'existe pas de production commerciale connue du BaP (EPA, 1990; IARC, 1983). La source anthropique principale du BaP est la combustion de bois de chauffage, les affleurements, les mines de charbon abandonnées, la fabrication de coke et la combustion de charbon bitumineux, la fusion de l'aluminium, la production de bitume et d'asphalte, les transports et l'industrie du pétrole et du gaz naturel (EPA, 1990; Health Canada., 1994). Le BaP peut être formé lors de l'utilisation ou la fabrication d'autres substances telles que les huiles de moteurs, l'essence, le tabac, les produits cosmétiques et pharmaceutiques, les huiles de cuisine, le beurre, la margarine, les aliments fumés, *etc.* (IARC, 1983). Le tableau 2-2 présente les émissions de BaP mesurées par Environnement Canada en 2012 dans la Province de Québec. Il a été constaté que l'industrie de l'aluminium est la plus importante source d'émission du BaP dans l'air.

Tableau 2-2 : Émissions de BaP (kg/année) dans la Province de Québec selon la base de données de l'Inventaire national des rejets de polluants en 2012 [a].

No INRP	Installation	Ville	Rejets sur place				Élimination		
			Air	Eau	Sol	Total	Sur place	Hors site	Total
3057	Rio Tinto Alcan - Usine Shawinigan	Shawinigan	2 114	0,100	0	2 115	0	1 515	1 515
3406	Rio Tinto Alcan - Usine Arvida	Saguenay	69	0,510	0	69	0	0	0
1872	Cascades Canada ULC. - NORAMPAC CABANO	Temiscouat a-sur-le- Lac	31	0	0	31	0	0	0
1071	Aluminerie de Bécancour Inc.	Bécancour	6,6	0	0	6,6	0	0	0

No INRP	Installation	Ville	Rejets sur place				Élimination		
			Air	Eau	Sol	Total	Sur place	Hors site	Total
3060	Rio Tinto Alcan - Usine Laterrière	Laterricre	1,4	0	0	1,4	0	0	0
5569	SGL Canada Inc.	Lachute	1,2	0	0	1,2	0	0	0
3928	Ultramar limitée - Raffinerie Jean-Gaulin	Lévis	0,861	0,018	0	0,879	0	1,2	1,2
3062	Rio Tinto Alcan - Usine Grande-Baie	La Baie	0,455	0	0	0,455	0	0	0
4778	Aluminerie Alouette inc.	Sept-Îles	0,360	0,018	0	0,378	0	0	0
2909	Stella-Jones Inc. - Usine de Delson	Delson	0,003	0	0	0,003	0	0,260	0,260
Total pour tous les installations			2225	0,646	0	2226	0	1 517	1517

[a] Données obtenues à partir du NPRI (2012).

2.1.4.3 Comportement dans l'environnement

Le comportement du BaP dans l'environnement est largement déterminé par son insolubilité dans l'eau et sa propension à se lier à d'autres particules ou d'autres substances organiques. Communément associé à la suie, le BaP dans l'atmosphère se trouve sous forme particulaire et finit par se déposer au sol (52% du BaP dans l'atmosphère finira par se déposer au sol). Une fois déposé, le BaP est capable de pénétrer facilement le sol. Dans l'eau, le BaP peut se lier aux particules en suspension ou être déposé dans les fonds sédimentaires. Le principal moyen de transport du BaP dans l'environnement est donc par voie particulaire dans l'air. La distribution environnementale du BaP dépend de la taille des particules présentes et des conditions météorologiques. Par conséquent, le BaP peut être transporté à de grandes distances de sa source d'émission (EPA, 1990). Dans l'atmosphère, le BaP peut subir des réactions chimiques avec l'ozone et le protoxyde d'azote ainsi que des réactions photochimiques, résultant en la formation d'espèces réactives oxygénées (Genium Publishing Corporation., 1999).

2.1.4.4 Exposition de la population générale

Comme exposé dans la section précédente, la plupart du BaP est présent dans l'air. Ainsi, l'exposition humaine peut se faire par inhalation. La population est susceptible d'être exposée par l'eau et la nourriture étant donné que plus de la moitié du BaP dans l'air finit par se déposer au sol et dans les systèmes aquatiques (ATSDR, 1995a). Les principales sources d'exposition résidentielles et commerciales sont dues au chauffage au bois, au charbon, à l'huile ou au gaz (voir tableau 2-3, Health Canada. (1994)).

Tableau 2-3 : Émissions annuelles des HAP au Canada selon le rapport sur les HAP d'Environnement Canada (Health Canada., 1994).

Sources	Émissions de HAP	
	Tonnes	%
Procédés industriels		
Usine d'aluminium	925	21
Usine métallurgiques	19,5	0,4
Production du coke	12,8	0,3
Raffineries de pétrole	2,5	0,1
Chauffage résidentiel		
Bois	474	11
Autres	29	0,7
Feux à ciel ouvert	358	8,3
Incinération		
Brûleur tipi	249	5,8
Municipale	1,3	<0,1
Industrielle	1,1	<0,1
Transports		
Diesel	155	3,6

9

Sources	Émissions de HAP	
	Tonnes	%
Essence	45	1
Autre	1,2	<0,1
Centrales thermoélectriques	11,3	0,3
Combustion industrielle		
Bois	5,7	<0,1
Autre	10,2	0,2
Chauffage commercial et institutionnel	2,7	0,1
Tabac	0,2	<0,1
Feux de forêts	2010	47
Total	**4314**	100

Par ailleurs, le mode de vie des gens peut aussi influencer leur exposition (par exemple, via le tabagisme et l'alimentation). À l'intérieur, la population est surtout exposée aux HAP suite à la cuisson des aliments et par la fumée de tabac. Il existe aussi des procédés de production qui sont des sources substantielles de BaP dans les aliments cuits sur le gril, trop grillés, rôtis, frits, fumés, certains céréales et grains (par certains procédés de séchage de grains) et les légumes cultivés dans des sols contaminés. À l'extérieur, la population est principalement exposée aux gaz d'échappements des véhicules, aux émissions industrielles et aux feux de forêts (EPA, 1990; EC, 2002; IARC, 2010). Les concentrations observées dans les milieux urbains sont de dix à cent fois plus élevées que celles mesurées en milieux ruraux (EPA, 1990).

2.1.4.5 Exposition au travail

Les industries de production d'aluminium sont celles présentant les émissions les plus élevées par année (tableau 2-2). Par conséquent, les travailleurs de ces industries sont plus à risque d'être exposés au BaP. Cependant, d'autres industries présentent aussi un risque d'exposition au travail. C'est le cas d'usines de liquéfaction et de gazéification du charbon, de production de coke et des fours à coke, de distillation de goudron de houille, de production de toitures et revêtements (brai de houille), d'imprégnation du bois à la créosote, de fabrication des électrodes de carbone, et électriques (EPA, 1990; EC, 2002; IARC, 2010). Les concentrations typiques de BaP dans l'air à proximité d'industries de l'aluminium peuvent atteindre les 100 μg/m^3. Des concentrations de BaP entre 10 et 20 μg/m^3 ont été mesurées lors de travaux de toiture et de pavage. Finalement, des concentrations de BaP de l'ordre de 1 μg/m^3 ont été observées lors d'activités de ramonage, de distillation de goudron de houille, de liquéfaction de charbon et dans des centrales électriques (IARC, 2010).

Les travailleurs sont exposés par inhalation au BaP dans leur milieu de travail. Cette voie peut être la voie principale d'exposition chez les travailleurs, surtout dans les cas impliquant des procédés de combustion où les travailleurs peuvent respirer la fumée produite. Dans certaines circonstances, la voie cutanée peut devenir la voie principale d'exposition au BaP en milieu de travail (Forster *et al.*, 2008).

2.2 La toxicologie du benzo(a)pyrène

Le Centre international de recherche sur le cancer (CIRC) estime avoir assez de preuves scientifiques sur la cancérogénicité chez l'animal (IARC, 2010) et le classe comme cancérogène reconnu pour l'être humain (groupe 1).

2.2.1 Cancérogénicité et génotoxicité

Actuellement, bien qu'il n'existe pas d'étude contrôlée sur la cancérogenèse du BaP chez l'humain, plusieurs données ont démontré des effets cancérogènes et mutagènes du BaP sur des souris, des rats, des lapins, des hamsters, des grenouilles, des poissons et des primates (ATSDR, 1995a; CIRC, 1983; EPA, 1990; EC, 2002).

Dans les principales études documentées chez les animaux, le BaP a été mis en cause dans la formation de tumeurs malignes:

i. du système respiratoire supérieur sur des hamsters mâles exposés par inhalation, principalement au nez, au larynx et au pharynx (Thyssen *et al.*, 1981);

ii. du système digestif supérieur sur des hamsters mâles exposés par inhalation (Thyssen *et al.*, 1981) et sur des souris femelles après une exposition alimentaire (Culp *et al.*, 1998);

iii. du système lymphatique (lymphomes) sur des souris exposées par injection intra-péritonéale (ATSDR, 1995a) et par voie orale (deVries *et al.*, 1997).

Thyssen *et al.* (1981) ont effectué une étude chronique pendant cent neuf semaines sur des hamsters exposés par inhalation. Ils ont trouvé des polypes, des papillomes et des carcinomes au tractus respiratoire. En particulier, la cavité nasale, le pharynx, le larynx, et la trachée ont développé des tumeurs en proportion à la dose administrée de BaP. Aucune tumeur n'a été détectée dans les poumons.

12

Également, dans une étude sous-chronique de vingt-six semaines sur des souris, l'administration intra-gastrique de 67-100 mg/kg de BaP a produit la formation de papillomes au niveau de l'estomac et d'adénomes pulmonaires (Sparnins *et al.*, 1986). L'expérience de Neal et Rigdon (1967) sur des souris a permis de relier la durée d'exposition avec l'incidence de carcinomes et papillomes dans l'estomac. Ces chercheurs ont administré par voie orale 33 mg/kg de BaP par jour pendant des périodes variant entre 30 et 197 jours (ATSDR, 1995b). Dans une autre étude d'exposition par voie orale faite sur des rats, l'incidence de tumeurs mammaires chez les rats traités était plus du double par rapport au groupe contrôle. Dans cette expérience, McCormick *et al.* (1981) ont administré 12,5 mg/kg de BaP pendant huit semaines et ont mesuré l'incidence de tumeurs quatre-vingt-dix semaines plus tard.

Tableau 2-4 : Quelques doses maximales sans effet nocif observable (DMSENO) et doses minimales pour un effet nocif observable (DMENO) pour les effets cancérogènes, issues d'études chroniques ou sous-chroniques chez l'animal.

	Période d'exposition	Espèce	Voie d'exposition	Dose	Effets	Références
DMSENO	4,5 heures/ jour, 5 jours/semain e pendant 16 semaines	Hamster	Inhalation	44,8 mg/m^3	Cancérogène : tumeurs aux poumons.	Thyssen *et al.* (1981)
DMENO	3-4,5 heures/ jour, 7 jours/semain e pendant 109 semaines	Hamster	Inhalation	9,5 mg/m^3	Cancérogène : augmentation de tumeurs au tractus respiratoire et néoplasmes dans la partie supérieure du tractus digestif.	Thyssen *et al.* (1981)
DMSENO	2 à 46 années	Travailleurs		5×10^{-4} mg/m^3	Génotoxique : augmentation de mutation de lymphocytes périphériques.	Perera et al. (1993)

Les seules observations d'effets cancérogènes du BaP chez l'être humain viennent des études réalisées chez des travailleurs exposés lors de leurs activités de travail. Or, ces travailleurs ne sont pas uniquement exposés au BaP. Ils sont surtout en contact avec des mélanges de HAP, et d'autres substances et particules. Ainsi, les effets toxiques résultants pourraient être intensifiés ou réduits par la co-exposition à d'autres substances (ATSDR, 1995a). Malgré cela, il y a suffisamment de preuves scientifiques, à partir d'études chez les travailleurs, pour conclure que le risque de cancer du poumon, de la vessie et de la peau est augmenté lorsqu'on accroît l'exposition cumulative au BaP (OIT, 1986; EPA, 1984).

Concernant le cancer du poumon, dans une étude de cohorte, le risque relatif a été respectivement évalué à 1,75 et 3,77 pour des expositions à des concentrations dans l'air de 10 et 480 $\mu g/m^3$ de BaP en moyenne par an chez les travailleurs de sept alumineries (Armstrong et Gibbs, 2009). Dans une expérience similaire, le risque relatif a été respectivement estimé à 1,23 et 1,79 pour une concentration d'exposition de 0,5-20 et >80 $\mu g/m^3$ de BaP par an chez les travailleurs d'une fonderie d'aluminium (Spinelli et al., 2006). Finalement, dans une usine d'acier, le risque relatif calculé à partir d'une étude cas-témoins a été 1,8 pour une concentration d'exposition dans l'air supérieure ou égale à 3,2 $\mu g/m^3$ de BaP par an (Xu et al., 1996). Il existe des études qui montrent une augmentation de la mortalité due au cancer du poumon chez des travailleurs exposés aux HAP (dont le BaP) dans la production de coke (Mazumdar et al., 1975 ; Redmond et al., 1976), ou affectés à la réfection de toiture en goudron (Hammond et al., 1976) et encore chez des individus exposés à des produits de combustion comme la cigarette (Maclure et MacMahon, 1980).

Les preuves scientifiques sont moins concluantes dans le cas du cancer de la vessie que dans le cas du cancer du poumon. Cependant, les expériences ont tout de même montré que le risque augmente avec le cumul d'exposition au BaP. Par exemple, dans une étude de cohorte chez les paveurs d'asphalte, des risques relatifs de 1,7 et 1,1 ont été respectivement calculés pour des concentrations de 0,3-0,9 et 0,9-1,7 $\mu g/m^3$ de BaP par an (Burstyn et al., 2007). Dans l'article de Spinelli et al. (2006), le risque relatif associé au cancer de la vessie

14

a été estimé à 1,2, 1,5 et 1,9 pour des concentrations variant respectivement entre 20 et 40, entre 40 et 80 et supérieures à 80 $\mu g/m^3$ de BaP par an.

En ce qui concerne le cancer de la peau, dans une étude de cohorte, Hammond *et al.* (1976) ont évalué le ratio standardisé de mortalité relatif pour le cancer de la peau (excluant les mélanomes) à 4,9 pour des travailleurs de l'asphalte pour une période d'activité variant entre 10 et 20 et à 4 pour une période d'activité supérieure à 20 ans. D'autres études associant le cancer de la peau à l'exposition au BaP ont estimé le ratio standardisé de mortalité (excluant les mélanomes) à 1,5 chez des travailleurs de l'industrie de l'énergie électrique lorsqu'ils étaient exposés à la créosote, au plomb, au silicium et aux isocyanates (Tornqvist *et al.*, 1986). Dans l'étude de cohorte de Karlehagen *et al.* (1992), le ratio standardisé d'incidence pour le cancer de la peau a été calculé à 2,4 (excluant les mélanomes) et à 1,7 (incluant seulement les mélanomes) pour des travailleurs exposés à l'huile de créosote.

2.2.2 Effets sub-aigus et aigus

Les études de toxicité aiguë décrivent les effets néfastes d'une substance qui résulte soit d'une exposition unique ou d'expositions répétées dans une courte période de temps (habituellement moins de 24 heures) et à des fortes doses d'administration (Hayes, 2008; Klaassen, 2001). Les effets toxiques aigus doivent se présenter dans les 14 jours suivant l'administration de la substance toxique.

2.2.2.1 Chez l'animal

Lors d'une exposition cutanée au BaP, les effets observés sont l'hypersensibilité au contact et l'immunosuppression de glandes sébacées (WHO, 1998). L'administration intra-péritonéale du BaP chez des souris a également mis en évidence des effets hématologiques, immunosuppresseurs, neurologiques (effets sur la production de certains neurotransmetteurs) et sur le système reproducteur (ATSDR, 1995a; EPA, 1990; WHO, 1998). Finalement, seulement une étude par inhalation aigüe a été documentée. Dans cette étude, les rats n'ont présenté aucune lésion dans les cavités nasales ni dans les poumons (Wolff *et al.*, 1989).

2.2.2.2 Chez l'humain

À notre connaissance, il n'y a pas d'étude publiée de toxicité aigüe ou sub-aigüe qui ait été réalisée sur des individus. Aucune voie d'exposition ni effet n'ont été étudiés ni documentés (Casarett *et al.*, 1996; CIRC, 1983; Klaassen, 2013).

2.2.3 Effets sous-chroniques et chroniques non cancérogènes

Les études de toxicité chronique ont pour objectif la caractérisation des effets toxiques sur un organisme vivant exposé à de faibles doses d'une substance toxique de façon continue (exposition prolongée) ou répétée (Hayes, 2008; Klaassen, 2013).

2.2.3.1 Chez l'animal

À part la cancérogénicité, l'exposition au BaP produit principalement des effets immunosuppresseurs, reproductifs et hématologiques. Il a été démontré que, chez des animaux, l'exposition *in vitro* et *in vivo* des lymphocytes à des doses élevées de BaP avait une incidence sur la production de cellules T (Blanton *et al.*, 1986; Wojdani et Alfred, 1984). Des effets immunosuppresseurs ont été observés par Ginsberg *et al.* (1989), accompagnés par des quantités élevées d'adduits du BaP à l'ADN. D'autre part, des études *in vitro* ont montré que le BaP ne semble pas causer d'effets immunologiques sans que des tumeurs ne soient également observées (Dean *et al.*, 1983).

Certaines expérimentations sur les animaux montrent des effets sur la reproduction chez des espèces exposées au BaP. Lors d'exposition orale de souris au BaP, Legraverend *et al.* (1984) ont montré que les métabolites produits par le fœtus sont responsables de la perte du fœtus, de malformations fœtales et de la stérilité de la progéniture. Différents types de dysmorphogenèse ont été observés dans l'étude d'exposition par voie intra-péritonéale de Shum *et al.* (1979) : des mort-nés, des pertes de fœtus et des malformations (pied bot, fente labiale ou palatine, queue bouclée, pigmentation anormale, *etc.*).

Shubik *et al.* (1960) ont également montré des effets hématologiques létaux après une seule dose intra-péritonéale de BaP sur des souris. Robinson *et al.* (1975) ont documenté que le décès des souris suite à une exposition par voie orale au BaP résulte d'une hémorragie, d'une anémie aplasique et/ou d'une pancytopénie.

Tableau 2-5 : Dose minimale avec effet nocif observé (DMENO) autre que le cancer, issues d'expériences chroniques ou sous-chroniques chez l'animal.

	Période d'exposition	Espèce	Voie d'exposition	Dose	Effets	Références
DMENO	6 mois	Souris	Orale	120 mg/kg par jour	Systémique: augmentation du poids du foie, dépression de la moelle osseuse et anémie aplasique	Robinson *et al.* (1975)

2.2.3.2 Chez l'humain

Les études d'exposition en milieu de travail à long terme sont la principale source d'information disponible sur la toxicité du BaP chez l'être humain. Il faut signaler que les études basées sur des données expérimentales collectées au travail peuvent difficilement estimer avec précision la dose d'exposition du BaP. De plus, le BaP est présent avec d'autres substances et d'autres matières particulaires. Ainsi, les effets toxicologiques observés pourraient être exacerbés par l'absorption de plusieurs types de particules. Dans une étude réalisée dans une usine de production de caoutchouc où plusieurs matières particulaires sont présentes, six cents soixante-sept travailleurs ont présenté une diminution des fonctions respiratoires (Gupta *et al.*, 1993). Nonobstant, cette observation peut difficilement être associée à une seule substance en particulier.

Quant à l'exposition cutanée, Cottini et Mazzone (1939) ont documenté les effets de l'application répétitive d'une solution à 1% de BaP. Dans leur étude de quatre mois, ils ont observé l'apparition des verrues bénignes. Bien que ces effets soient réversibles, ils ont présumé que ceux-ci étaient le résultat d'une prolifération néoplasique. Quant aux effets immunologiques, une suppression immunitaire humorale a été documentée chez des travailleurs de fonderie de fer en Pologne (Szczeklik *et al.*, 1994).

18

Tableau 2-6 : Quelques doses maximales sans effet nocif observable (DMSENO) et doses minimales avec effet nocif observé (DMENO) autre que le cancer, issues d'études chroniques ou sous-chroniques par inhalation chez l'humain.

	Période d'exposition	Concentration de BaP dans l'air	Effets	Références
DMSENO	0,5 à 6 années	1×10^{-4} mg/m^3	Systémique : toux, vomissement du sang, irritation de la gorge et poitrine, fonctions pulmonaires réduites.	Gupta *et al.* (1993)
DMSENO		0.2-0.4 mg/m^3	Génotoxique : adduits au ADN de placentas.	Perera *et al.* (1988)
	15 années en moyen	$2\text{-}5000\times10^{-4}$ mg/m^3	Immunologique : immunoglobuline sérique réduite.	Szczeklik *et al.* (1994)

2.3 La toxicocinétique du BaP

Le BaP est facilement absorbé dans l'organisme animal et humain suivant différents voies d'exposition : l'inhalation, l'ingestion et le contact cutané. Les expérimentations animales montrent également que la vitesse d'absorption ainsi que la fraction absorbée du BaP dépendent du véhicule utilisé pour l'administration. Le BaP rejoint ensuite la circulation systémique où il est rapidement distribué dans les organes. Selon les données disponibles chez l'animal, la biotransformation du BaP se fait principalement dans le foie mais les sites d'absorption peuvent aussi contribuer à la métabolisation du BaP. Seulement dix pour cent de cette substance est directement éliminé par voie fécale et urinaire étant donné que la vitesse de biotransformation du BaP est très grande.

2.3.1 Absorption

Le BaP est une substance lipophile (très peu soluble dans l'eau, EC, 2002). Ainsi, pour la voie respiratoire, Sun *et al.* (1982) ont exposé des rats à un aérosol du BaP pur et à un aérosol du BaP adsorbé sur des particules ultrafines (0,1 μm) d'oxyde de gallium. La fraction absorbée par les poumons après trente minutes étaient autour de 10% pour l'aérosol de BaP pur et de seulement 3% pour le BaP adsorbé à la surface de particules ultrafines. Ces mêmes chercheurs ont observé que l'entièreté du BaP inhalé rejoignait la circulation sanguine systémique, tandis que le BaP adsorbé sur des particules ultrafines était retenu par les poumons. Jusqu'au présent, il n'existe pas d'étude sur l'absorption du BaP par inhalation chez l'humain.

L'absorption du BaP par voie orale est rapide et présente un comportement exponentiel (Rees *et al.*, 1971). Dans une expérience sur des cochons cathétérisés, Laurent *et al.* (2001) ont mesuré la radioactivité du BaP marqué au ^{14}C dans la veine porte dans l'heure qui suivit l'administration orale. La radioactivité a atteint sa valeur maximale après cinq heures. Rees *et al.* (1971) ont mesuré de grandes concentrations de BaP dans les ganglions lymphatiques thoraciques trois heures après l'exposition intra-gastrique du BaP. Le tractus gastro-intestinal peut également contribuer à la biotransformation du BaP et les métabolites produits peuvent facilement traverser les parois intestinales pour atteindre la circulation sanguine. Dans une expérience sur des rats, Kawamura *et al.* (1988) ont démontré que la composition des aliments avec lesquels le BaP était administré avait un grand impact sur son absorption. Ainsi, lorsque le BaP était administré avec des aliments lipophiles, entre 42 et 50% du BaP était absorbé (en référence avec les résultats d'une injection intraveineuse). Pour d'autres types d'aliments, la proportion absorbée diminuait entre 20 et 30%. À notre connaissance, il n'y a pas eu d'expérience par voie orale chez l'humain.

20

L'absorption du BaP par voie cutanée est relativement rapide et la biotransformation *in situ* représenterait un rôle important lors du processus d'absorption selon certains auteurs (Brinkmann *et al.*, 2013; Jacques *et al.*, 2010; Kao *et al.*, 1985; Ng *et al.*, 1992; Storm *et al.*, 1990). Dans une étude sur des souris, Sanders *et al.* (1984) ont observé l'absorption de 6% du BaP marqué au ^{14}C dans l'heure qui a suivi l'application. Dans une expérience *in vitro* sur la pénétration du BaP, Kao *et al.* (1985) ont trouvé des différences dans l'absorption du BaP au niveau de la peau dorsale de plusieurs espèces animales. Après vingt-quatre heures, la peau de la souris a permis une pénétration de 10%. L'absorption du BaP à travers la peau du ouistiti, du rat et du lapin était de l'ordre de 1 à 3% tandis que celle du cochon d'Inde a seulement été de 0,1%. Dans le même article, les auteurs ont évalué la pénétration du BaP à travers la peau humaine à 3%. Dans une autre étude sur l'absorption par voie cutanée *in vitro*, Moody *et al.* (2007) ont observé une absorption de 56% de BaP marqué au ^{14}C en solution dans l'acétone par la peau humaine, tandis que seulement 15% a été absorbé lorsque la peau était exposée au sol de jardinage commercial enrichi au BaP marqué au ^{14}C.

2.3.2 Distribution

Le BaP absorbé est rapidement distribué dans les tissus (voir tableau 2-7). Or, la quantité de BaP distribué dans chaque tissu dépend de la voie et du véhicule d'administration (EPA, 1990). Peu importe la voie d'administration, des niveaux détectables de BaP sont retrouvés dans tous les organes et tissus entre quelques minutes et l'heure suivant l'administration (voir, par exemple, Moir *et al.* (1998)). Comme le BaP est une substance liposoluble, il est largement distribué dans les tissus adipeux et les glandes mammaires (Agbato, 2006). Des quantités élevées de BaP ont été également trouvées dans le foie (Foth *et al.*, 1988). La distribution importante du BaP dans le foie facilite la biotransformation par cet organe. Par ailleurs, il a été documenté que le BaP peut facilement traverser aussi la barrière placentaire chez le rat et la souris (Madhavan et Naidu, 2000; Neubert et Tapken, 1988; Withey *et al.*, 1993).

Dans une étude par voie intra-trachéale, Weyand et Bevan (1986) ont trouvé que la plupart du BaP marqué au ^{3}H se distribue rapidement aux poumons (60%) et dans le foie (13%) cinq minutes après l'administration et qu'environ 14% de la radioactivité étaient quantifiables dans la carcasse à ce temps. Après une heure, la radioactivité a été mesurée principalement dans la carcasse (27%), dans les intestins (24%), dans le foie (16%) et dans les poumons (15%).

Tableau 2-7 : Distribution du BaP en pourcentage de la dose administrée après administration intraveineuse de 40 µmol/kg de BaP sur des rats Sprague-Dawley. Adapté de Marie *et al.* (2010).

Tissus	Pourcentage de la dose administrée retrouvée sous forme de BaP							
	Temps après l'injection intraveineuse (h)							
	2	4	8	16	24	33	48	72
Sang	0,4	0,1	0,1	0,0	0,0	0,0	0,0	0,0
Reins	0,5	0,2	0,2	0,0	0,0	0,0	0,0	0,0
Foie	2,3	1,9	1,8	0,5	0,2	0,2	0,1	0,0
Poumons	17,0	10,2	17,6	16,9	8,0	11,2	7,4	5,2
Tissus adipeux	2,5	1,4	3,2	3,2	2,1	2,4	1,3	0,7
Peau	7,6	2,7	1,7	0,8	0,1	0,0	0,0	0,0
Fèces	0,0	0,0	0,0	0,1	0,4	0,4	0,4	0,4
Total	30,3	16,5	24,6	21,6	10,8	14,2	9,1	6,3

Pour la voie orale, Ramesh *et al.* (2001b) ont documenté que la cinétique de distribution du BaP était de premier ordre. Dans leur étude, ils ont observé une grande distribution du BaP dans le plasma sanguin six heures après l'administration du BaP. Après vingt-quatre heures, 10% de la dose se trouvait dans le foie tandis que 45% avait déjà été excrété dans les urines et fèces.

Tableau 2-8 : Distribution du BaP marqué au ^{3}H en pourcentage de la dose administrée après administration de 4 pmol/kg de BaP par instillation intra-trachéale à travers une canule trachéale chez des rats Sprague-Dawley. Adapté de la publication de Weyand et Bevan (1986).

Tissus	Pourcentage de la dose administrée retrouvée sous forme de radioactivité							
	Temps après l'application (min)							
	5	10	15	30	60	90	120	360
Sang	3,9	3,0	3,8	1,6	1,6	2,1	1,9	1,7
Reins	1,1	1,9	1,9	1,4	2,2	2,4	2,0	2,0
Foie	12,5	20,8	19,6	18,6	15,8	17,1	13,2	4,6
Poumons	59,5	41,9	32,2	20,4	15,4	11,3	11,0	5,2
Intestines	1,9	3,3	4,9	5,2	9,9	12,6	11,0	14,9
Contenu intestinal	0,5	0,8	2,9	3,3	14,3	17,1	16,8	44,7
Carcasse	14,4	26,4	23,0	26,3	27,1	25,0	22,3	21,5

La distribution par voie sous-cutanée a été jugée similaire à celle par administration intraveineuse et intra-trachéale sur des souris et des rats (WHO, 2000).

2.3.3 Biotransformation

Le BaP est principalement métabolisé par le foie, notablement par les cytochromes P450 1A1, 1A2 et 1B1 (Foth *et al.*, 1988; Gelboin, 1980). Cependant, il peut être également biotransformé par d'autres tissus tels que la peau (Fox *et al.*, 1975), les poumons (Mehta *et al.*, 1979) et le tractus gastro-intestinal (Autrup *et al.*, 1982; Harris *et al.*, 1979). Le système d'oxydases à fonction mixte du cytochrome P-450 est fortement impliqué dans la transformation du BaP (Lee *et al.*, 2003; Wood *et al.*, 1976). Dans une première étape (phase I) impliquant des processus oxydatifs et hydrolytiques, le BaP est activé par les cytochromes CYP1A1 et CYP1B1 (Kim *et al.*, 1998). Par la suite (phase II), ces dérivés réactifs sont conjugués aux acides glucuroniques, aux sulfates et au glutathion pour leur élimination (EPA, 1990).

Tableau 2-9 : Métabolisme du BaP par des enzymes P450 humaines tel qu'adapté de Kim *et al.* (1998).

Métabolite	Pourcentage de métabolites formés (%)		
	CYP1A1	CYP1A2	CYP1B1
9,10-diolBaP	12,8	<3,3	10,3
7,8-diolBaP	15,5	<3,3	16,8
BaP-1,6-quinone	11,4	16,7	13,3
BaP-3,6-quinone	19,8	20	15,9
BaP-6,12-quinone	8,1	25	16,8

	Pourcentage de métabolites formés (%)		
Métabolite	CYP1A1	CYP1A2	CYP1B1
9-OHBaP	14,1	11,7	13,1
3-OHBaP+7-OHBaP	18,2	20	16,8

Comme le montre la figure 2-1, le BaP est rapidement converti en une multitude d'époxydes par l'incorporation d'un groupement hydroxyle à la molécule (Angerer et Deutsche Forschungsgemeinschaft., 2006). Par la suite, les époxydes formés peuvent subir, à leur tour, une panoplie de réactions conduisant à une gamme très complexe et riche de sous-produits (Ariese *et al.*, 1994). Parmi ces réactions, l'hydrolyse des époxydes par l'époxyde hydrolase conduisant à des dihydrodiols et les réarrangements spontanés de phénols constituent les voies les plus importantes dans la production des sous-produits du BaP (Jongeneelen *et al.*, 1987). Ces deux voies finiront par conduire à des molécules mutagènes ou cancérogènes telles que le BaP-7,8-diol-9,10-époxyde (Anakwe *et al.*, 2002).

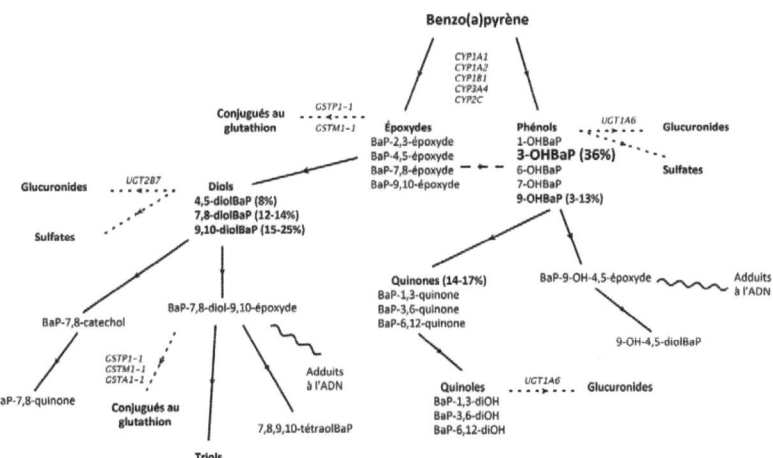

Figure 2-1: Voies métaboliques du BaP et de ses métabolites représentatifs tel que proposé par la Commission Européenne (2002).

En ce qui concerne la voie de formation des phénols, Shu et Bymun (1983) signalent une augmentation proportionnelle de l'excrétion des métabolites en fonction du degré d'hydroxylation. Au niveau de la voie de formation des phénols, on retrouve le métabolite 3-OHBaP. Dans une étude réalisée sur les cellules hépatiques, Yang *et al.* (1975) ont constaté que le 3-OHBaP est un métabolite majeur chez les rats, comme les pourcentages relatifs des métabolites le montrent: 3-OHBaP, 36% ; 9,10-diolBaP, 15% à 25%; BaP-quinones, 14% à 17% ; 7,8-diolBaP, 12% à 14%; 9-OHBaP, 3% à 13%; 4,5-diolBaP, 8%. Les pourcentages relatifs de métabolites produits par les enzymes cytochromes P450 humaines sont présentés au tableau 2-9.

Rey-Salgueiro *et al.* (2008) ont mesuré du 3-OHBaP non conjugué ou sous forme conjuguée aux glucuronides ou aux sulfates dans le tractus intestinal de différentes espèces (vache, cheval, lapin et porc). Ces animaux ont été nourris avec des aliments possiblement contaminés aux HAP. Finalement, Yang et ses collaborateurs ont mis en évidence une sulfatation et une glucuronidation du 3-OHBaP dans les tissus aortiques aviaires et dans les cellules endothéliales cultivées (Yang *et al.*, 1986a; Yang *et al.*, 1986b).

2.3.4 Excrétion

Au début des années quatre-vingt, Chipman et ses collaborateurs ont démontré que l'excrétion biliaire était la principale voie d'élimination du BaP, indépendamment de la voie d'administration. Dans leurs études sur des rats dont le conduit biliaire a été canulé, ils ont observé une excrétion biliaire de 60% et une excrétion urinaire de 3% du BaP marqué au [14]C

26

six heures après une administration intraveineuse (Chipman *et al.*, 1982; Chipman *et al.*, 1981a; Chipman *et al.*, 1981b; Chipman *et al.*, 1981c). Étant donné son caractère lipophile et selon les expérimentations animales réalisées, il a été démontré que le BaP est métabolisé avant d'être excrété (Likhachev *et al.*, 1992). Le tractus gastro-intestinal peut contenir des niveaux relativement élevés de métabolites résultant directement de la sécrétion biliaire à partir du foie (Wiersma et Roth, 1983). Chipman et ses collaborateurs ont également suggéré que l'élimination du BaP suite à l'exposition par voie orale serait plus rapide puisque l'effet du premier passage dans le foie aurait une contribution importante par rapport aux autres voies d'exposition. Auparavant, il avait déjà été démontré que le taux d'élimination était limité par le taux de métabolisme et non pas par le taux d'excrétion biliaire (EPA, 1990; Schlede *et al.*, 1970). Ainsi, l'élimination fécale du BaP suite à son ingestion serait plus rapide que lors d'une absorption par inhalation ou par contact cutané (EPA, 1990).

Bouchard et Viau (1997) ont analysé l'excrétion urinaire du BaP par voie orale, intraveineuse et cutanée chez le rat. Dans leur étude, ils ont observé une élimination urinaire des métabolites principaux du même ordre de grandeur pour les trois voies d'exposition et pour quatre doses différentes. Quant au BaP lui-même, la plupart des auteurs n'ont trouvé que des traces de cette substance. Ainsi, Cao *et al.* (2005) ont documenté que 0,00012% du BaP était excrété dans l'urine de rats dans les vingt-quatre heures après une exposition par voie intraveineuse. Lee *et al.* (2003) ont rapporté que 0,0065% de la dose du BaP se retrouvait dans l'urine de rats au cours des quatre-vingt-seize heures après une exposition par voie intrapéritonéale tandis que Marie *et al.* (2010) n'ont pas détecté de BaP non métabolisé. Finalement, Moir *et al.* (1998) ont documenté une excrétion urinaire de 6,2, 5,7 et 7,2% trente-six heures après une exposition par voie intraveineuse de 2, 6 et 15 mg/kg de BaP marqué au ^{14}C chez le rat.

2.4 Valeurs toxicologiques de référence et normes

Lors de la dernière évaluation en 2011, le BaP a été classé à la huitième place de la *liste prioritaire des substances dangereuses* de l'Agence pour les substances toxiques et le registre de maladies des États-Unis (ATSDR, 2011), d'où l'importance d'établir des recommandations et des normes concernant l'exposition au BaP. Le tableau 2-10 résume l'ensemble des valeurs toxicologiques de référence pertinentes à l'exposition au BaP par inhalation, ingestion et contact cutané.

Tableau 2-10 : Valeurs toxicologiques de référence relatives au BaP.

Norme réglementaire ou *niveau recommandé*	Valeur	Commentaires	Référence
	Inhalation		
Valeur de référence	2×10^{-6} mg/m³	Valeur obtenue sur la base d'effets non cancérogènes (diminution de la survie du fœtus).	EPA (1990)
Limite admissible pour 8 heures d'exposition moyenne pondérée	0,2 mg/m³	Fraction soluble dans le benzène de goudron de houille de brai volatil.	EPA (1990)
Valeur limite de seuil pour 10 heures d'exposition moyenne pondérée	0,1 mg/m³	Fraction soluble dans le cyclohexane de goudron de houille de brai volatil.	EPA (1990)
Limite admissible pour 8 heures	0,2 µg/m³	BaP.	EPA (1990)

Norme réglementaire ou *niveau recommandé*	Valeur	Commentaires	Référence
d'exposition moyenne pondérée			
Excès de risque unitaire	**5×10^{-4} par μg/m^3**	Facteur basé sur des tumeurs au tractus respiratoire et gastro-intestinaux de hamsters.	EPA (1990)
	Ingestion		
Valeur de référence	**3×10^{-4} mg/kg par jour**	Valeur obtenue sur la base d'effets non cancérogènes (principalement, reproductifs et immunologiques).	EPA (1990)
Critère de qualité de l'eau ambiante	0 (28, 2,8 et 0,28 ng/L)	HAP totaux (selon l'estimation de l'excès de risque à 10^{-5}, 10^{-6} et 10^{-7}).	EPA (1990)
Niveau maximum de contaminant pour l'eau potable	2 μg/L	Les critères étant les difficultés reproductives et l'augmentation du risque de cancer.	WHO (2000)
Excès de risque unitaire	**1×10^{-3} par μg/kg par jour**	Facteur basé sur des tumeurs aux cellules squameuses de l'estomac de souris et de rats.	EPA (1990)
Excès de risque unitaire	7,3 par mg/kg par jour	Facteur basé sur des tumeurs dans le tractus gastro-intestinal de rats.	WHO (2000)
	Cutanée		
Excès de risque unitaire	**5×10^{-3} par μg par jour**	Facteur basé sur des données de souris.	EPA (1990)

2.5 Évaluation de l'exposition

L'évaluation de l'exposition humaine aux produits toxiques est un processus essentiel en santé publique lorsqu'il s'agit d'identifier, de prévenir et de contrôler l'exposition des populations à des produits chimiques potentiellement dangereux. Il existe deux approches principales en évaluation de l'exposition humaine (Spengler *et al.*, 2000).

D'une part, la surveillance environnementale permet d'obtenir des résultats en référence à l'environnement auquel un groupe de la population est exposé. D'autre part, la surveillance biologique évalue directement les quantités de produit toxique présentes dans l'organisme exposé. Dans les deux cas, il est important de posséder une connaissance approfondie de la relation entre la concentration mesurée du xénobiotique et la réponse toxique observée.

2.5.1 Les méthodes d'évaluation de l'exposition

Dans le cadre du présent travail, nous nous sommes intéressés aux aspects concernant l'évaluation de l'exposition chez des travailleurs. Cette évaluation ne doit pas inclure seulement la quantification de la substance toxique, elle doit également considérer d'autres facteurs caractérisant l'exposition à un composé toxique tels que : le taux de production, l'intensité, la fréquence, la durée, la voie d'absorption, le taux d'absorption, *etc.* (Nielsen *et al.*, 2008). Les milieux de travail sont des environnements dynamiques qui sont constamment en changement. Ainsi, les travailleurs sont souvent exposés à de nouvelles substances, de nouveaux procédés, des horaires de travail variables, et il y a des permutations d'employés. Par conséquent, les scénarios d'exposition varient constamment. Ceci rend très difficile la caractérisation exacte de l'exposition en fonction du temps (Leeuwen et Vermeire, 2007).

Actuellement, il existe deux types de méthodes principalement utilisées dans l'évaluation de l'exposition. Une approche de mesure à la source de l'exposition (surveillance environnementale) et une autre approche de mesures centrées directement sur l'individu concerné (surveillance biologique). Dans les sections suivantes, les deux techniques sont décrites brièvement.

30

2.5.1.1 Surveillance environnementale

Dans la première approche d'évaluation de l'exposition, il faut rassembler toute l'information sur les sources et leurs facteurs déterminants, spécifiques et caractéristiques du milieu d'exposition qui peuvent agir sur l'être l'humain. La mesure des concentrations de BaP dans l'air contribue à estimer l'importance de l'exposition. Ainsi, des mesures de cette concentration peuvent servir pour établir le risque, sachant que la dose minimale avec effet nocif observé chez des hamsters est de 9,5 mg/m^3 (Thyssen *et al.*, 1981). Cette technique permet aux toxicologues de focaliser leur attention sur les sources de produits toxiques et de chercher à réduire leur impact.

2.5.1.2 Surveillance biologique

Pour la surveillance biologique, la connaissance des facteurs déterminants est aussi importante que dans l'approche de surveillance environnementale sauf que l'accent est mis sur l'impact de différentes sources de substances toxiques pour chaque individu (OIT, 1986; Que Hee, 1993). Ainsi, des mesures de biomarqueurs sont nécessaires pour chaque travailleur. Ces mesures sont ensuite comparées aux niveaux de biomarqueurs permissibles. En fonction de la durée de l'exposition et du produit toxique, le niveau de biomarqueurs varie dans les différents échantillons biologiques accessibles (voir tableau 2-11). En général, un biomarqueur est soit une mesure de la molécule toxique dans les matrices biologiques, soit un indicateur biochimique, génétique ou moléculaire qui sert à dépister la toxicité d'une substance (Hayes, 2008). On parle ainsi respectivement de biomarqueur d'exposition et de biomarqueur d'effet précoce. En fonction de l'affinité des substances pour les organes et de la vitesse à laquelle la substance toxique est métabolisée et éliminée de l'organisme, elle peut se retrouver à divers sites peu de temps après son absorption et à d'autres sites après

31

une plus longue période de temps. Toutefois, afin de déterminer la relation entre la dose d'exposition de la substance toxique étudiée et les effets analysés, une connaissance approfondie de la cinétique de la substance est nécessaire. Ainsi, la surveillance biologique est une technique fondamentalement nécessaire dans l'évaluation de l'exposition et la caractérisation du risque toxique (Nielsen *et al.*, 2008).

Tableau 2-11 : Matrices biologiques à priviligier selon le contaminant d'intérêt et le type d'exposition (adapté de Tardiff et Goldstein (1991)).

Substance	Exposition récente	Exposition à long terme
Aluminium	Plasma	Os
Arsenic	Urine	
Cadmium	Sang	Reins, foie et urine
Chrome	Urine et plasma	Erythrocytes
Cuivre	Sérum et plasma	
Plomb	Sang	Dents et os
Mercure	Sang et urine	Cheveux
Nickel	Urine et plasma	
Sélénium	Urine	
Nicotine	Sang et plasma	
Benzène	Air exhalé	Tissus adipeux

De cette manière, l'attention est focalisée sur les travailleurs de façon individuelle. L'évaluation du risque est donc personnalisée et tient compte de l'exposition totale du travailleur. Ainsi, l'évaluation inclut l'exposition au travail, l'exposition par l'alimentation, les déplacements du travailleur, le mode de vie et toute autre activité pouvant affecter l'exposition du travailleur en dehors de son milieu de travail. Dans le cas du BaP, un travailleur moyennement exposé au travail pourrait être plus à risque que ses collègues s'il est aussi exposé au BaP par la nourriture ou la cigarette (IARC, 1983). Par conséquent, cette méthode de surveillance biologique est un peu plus avantageuse par son approche personnalisée et globale. Cependant, cette technique porte également ses désavantages. Par exemple, des prélévements biologiques sont nécessaires ainsi que la participation active des travailleurs. Malgré cela, la détermination de l'exposition par mesure directe chez les travailleurs reste une méthode très fiable et celle-ci est la technique préférentiellement utilisée pour la surveillance biologique de l'exposition aux HAP (Lafontaine *et al.*, 2004). Sachant que les HAP, et le BaP en particulier, sont des produits chimiques dont les concentrations observées ne doivent jamais dépasser le niveau d'exposition recommandé par les organismes réglementaires, il devient très important pour les groupes à risque d'être sous surveillance biologique régulière (Angerer et Deutsche Forschungsgemeinschaft., 2006; Knudsen *et al.*, 2011a; Knudsen *et al.*, 2011b).

2.5.2 La surveillance biologique de l'exposition au BaP

Les méthodes de surveillance biologique des HAP emploient plusieurs types de biomarqueurs. Les plus couramment utilisés ont été la mesure des HAP et de leurs métabolites dans le sang et l'urine, la mesure du pouvoir mutagène dans l'urine et les fèces à travers la détermination des aberrations chromatiques, des échanges de chromatides sœurs dans les lymphocytes du sang, et des adduits à l'ADN dans plusieurs tissus (WHO, 2000). Plus particulièrement, le 1-hydroxypyrène (1-OHP, un métabolite du pyrène) a été utilisé comme biomarqueur de préférence pour la surveillance biologique de l'exposition aux HAP (ATSDR, 1995a). Ainsi, grâce à la détection du 1-OHP chez dix-neuf travailleurs d'une usine de production de pierre réfractaire, Gundel *et al.* (2000) ont observé une exposition élevée à divers HAP (probablement due à une absorption cutanée). Ces chercheurs ont pu détecter également le 3-OHBaP et ils ont souligné qu'il devrait davantage être étudié, car ce métabolite est corrélé au potentiel cancérogène des HAP. Ils ont proposé l'utilisation du 3-OHBaP comme biomarqueur de l'exposition aux HAP cancérogènes afin de prévenir les risques cancérogènes résultant de l'exposition au BaP.

33

Plus récemment, Campo *et al.* (2010) ont évalué l'exposition aux HAP de cinquante-cinq travailleurs de fours à coke en déterminant les profils urinaires du 1-OHP et du 3-OHBaP (parmi d'autres HAP hydroxylées et non-métabolisés). Ils ont trouvé que les mesures de ces deux métabolites dans les urines étaient des biomarqueurs utiles pour estimer l'exposition aux HAP. Auparavant, une corrélation positive entre le 3-OHBaP et 1-OHP dans l'urine avait déjà été observée par Jongeneelen *et al.* (1986) dans une étude sur cinq patients subissant des traitements dermatologiques à base de goudron de houille. Il y a quelques années, Forster *et al.* (2008) ont évalué l'exposition aux HAP de deux cent cinquante-cinq travailleurs de différents secteurs d'activité. Le BaP a été détecté dans tous les lieux de travail et le 3-OHBaP a été retrouvé dans les urines analysées. En outre, des corrélations positives entre le 3-OHBaP et le 1-OHP dans l'urine ont été observées chez les travailleurs de cokeries et de production d'électrodes en graphite confirmant la fiabilité de 3-OHBaP comme biomarqueur au même titre que le 1-OHP.

Enfin, Lafontaine *et al.* (2004) avaient proposé une valeur de référence préliminaire de 0,4 nmol/mol de créatinine de 3-OHBaP dans un prélèvement urinaire obtenu 24 heures après le début de l'exposition. Cette valeur correspond à la valeur guide de BaP dans l'air de 150 ng/m^3 proposée par la Caisse nationale de l'assurance maladie française (Lafontaine *et al.*, 2004).

2.6 Modélisation toxicocinétique

Récemment, des efforts considérables ont été déployés, dans les études d'évaluation des risques, afin de mettre en place des outils fiables permettant de reconstituer les doses d'exposition d'individus à partir de mesures de biomarqueurs d'exposition retrouvés dans des

34

échantillons biologiques accessibles tels que l'urine, la salive, le sang, *etc.* (McNally *et al.*, 2012 ; Bouchard et al., 2001, 2006 ; Carrier et al., 2001). Dans de nombreuses études, il a été établi qu'une évaluation de l'exposition professionnelle à des substances chimiques peut être effectuée par la mesure de biomarqueurs d'exposition à différents moments dans le temps (Berthet *et al.*, 2012; Bouchard *et al.*, 2010; Gosselin *et al.*, 2006; Gosselin *et al.*, 2005; Heredia-Ortiz *et al.*, 2011; Heredia-Ortiz et Bouchard, 2012; McNally *et al.*, 2012; McNally *et al.*, 2011). Pour y parvenir, un modèle mathématique peut être construit afin de déterminer la relation entre les concentrations d'exposition d'un produit toxique et ses biomarqueurs en fonction du temps. Ce type de modèle permet de décrire les déterminants essentiels de la cinétique et peut alors être utilisé pour reconstruire les doses absorbées à partir des profils temporels de biomarqueurs chez les individus exposés.

Au début des années 1930, des modèles à compartiments séquentiels ont été développés pour simuler des données expérimentales en pharmacologie. C'est seulement dans les années 1960 que ce type de modèles a été reconnu utile en raison des nouvelles connaissances sur la relation entre la dose et l'effet observé et l'importance des concepts cinétiques tels que la clairance et les phénomènes dépendants du débit cardiaque. Depuis les années 1990, l'application des modèles cinétiques a connu une expansion importante dans l'évaluation des risques toxicologiques (Wagner, 1981 ; Krishnan et Andersen, 2010; Rescigno, 2010; Gerlowski et Jain, 1983; Lipscomb et Ohanian, 2007; OIT, 1986).

Le défi principal de la modélisation toxicocinétique est de réussir à créer une représentation fidèle de la physiologie et du devenir du composé d'intérêt et de ses principaux métabolites dans le sang, les tissus clés et les excrétas. Généralement, en toxicologie, le terme *cinétique* décrit le cheminement d'un composé toxique dans l'organisme, incluant l'absorption, la distribution, le métabolisme et l'excrétion (ADME), Dans certains cas, nous pouvons être intéressés par les caractéristiques spécifiques de la distribution aux organes et

35

par la métabolisation du produit (A**DM**E) dans l'organisme (Peters, 2011). Dans d'autres cas, nous pouvons être plutôt intéressés par l'absorption et l'élimination globale (**A**DM**E**) de la substance toxique et de ses dérivés dans l'ensemble de l'organisme (Atkins, 1969; Berg, 2011; Rubinow, 2002).

Dans le premier cas, la connaissance détaillée de la toxicocinétique du xénobiotique et de ses métabolites est indispensable pour estimer les concentrations aux organes cibles donc les effets biologiques en découlant (soit la toxicodynamique). La modélisation focalise alors sur la distribution et la métabolisation (A**DM**E) de la substance toxique. Dans le deuxième cas, la connaissance des vitesses d'absorption et d'élimination globale de la substance et de ses métabolites de l'organisme est nécessaire afin de relier les excrétions observées à des profils d'exposition (**A**DM**E**). Par vitesse d'élimination, on réfère à l'élimination globale d'un produit de l'organisme sans se soucier des détails du devenir dans l'organisme.

2.6.1 Modèles à compartiments

Bien que la modélisation mathématique dans les sciences de la santé ait une longue histoire (voir, par exemple, les travaux épidémiologiques sur la pratique de la vaccination contre la variole de Daniel Bernoulli autour de 1760, Brauer (2008)), le concept moderne de modélisation compartimentale sur lequel repose la modélisation actuelle a été introduit par les travaux de Rutherford et Soddy au début du 20ᵉ siècle (Rutherford, 1962). Dans leur étude, ils ont décrit la désintégration radioactive du thorium par une série de compartiments où le taux de désintégration était proportionnel au nombre d'atomes présents. Plus tard, en 1924, Widmark et Tandberg ont publié leurs travaux sur un modèle à un compartiment par administration intraveineuse, pour décrire l'élimination sanguine de plusieurs narcotiques (tel que cités dans Wagner (1981)). En 1935, le même comportement exponentiel a été observé par Behnke et ses collaborateurs lors de la modélisation à deux compartiments en parallèle pour décrire l'élimination de l'azote par des chiens et des êtres humains (tel que cité par Rescigno (2010)).

Au départ, le but primaire de la modélisation était de reproduire les observations expérimentales ; la modélisation était alors limitée par les connaissances disponibles sur la toxicocinétique et les expériences réalisables en laboratoire. À défaut de connaître la cinétique détaillée (absorption, distribution, métabolisme et excrétion), les premiers types de modèles furent donc construits pour représenter la cinétique globale, c'est-à-dire que l'organisme était modélisé en fonction des caractéristiques d'entrée et de sortie de la molécule d'intérêt (**ADME**) sans connaissance de son cheminement détaillé dans l'organisme (modèle de boîte noire, voir par exemple Leonov (2001)). Par la suite, lorsque la connaissance des processus biologiques et physiologiques s'est approfondie, les modèles ont eu tendance à se complexifier pour simuler tous les éléments essentiels de distribution et de métabolisation (A**DM**E). Ces modèles physiologiques seront présentés dans la section 2.6.2 mais, avant, les principales caractéristiques des modèles à compartiments simples seront décrits.

2.6.1.1 Représentation des compartiments

Typiquement, dans les modèles toxicocinétiques à compartiments, les compartiments peuvent représenter des organes individuels, des ensembles d'organes ou des fonctions (Aiache, 1985; Atkins, 1969; Jacquez, 1985). Le cas le plus simple consiste à modéliser l'ensemble de l'organisme par un compartiment unique tel que montré à la figure 2-2. Dans cette figure, la ligne pointillée représente l'entrée de la substance toxique. Le profil temporel

D(t) représente la dose administrée en fonction du temps. Les lignes en trait continu représentent la sortie de la substance de l'organisme avec un taux d'élimination k. Le cercle noir représente la quantité du xénobiotique B(t) dans l'organisme (seul compartiment, communément représenté par un rectangle). Les cercles blancs marqués d'un x représentent les sites de sortie de la substance toxique (typiquement non représentés, on devrait cependant les considérer aussi comme des compartiments). Dans ce cas particulier, nous avons différencié les quantités de la substance éliminée par voie urinaire U(t) et par d'autres voies d'élimination O(t). En pratique, il est plus facile de collecter la substance éliminée par une voie particulière (urine, fèces, cheveux, sueur, salive) plutôt que de collecter la totalité du xénobiotique sortant du corps. Dans la figure, α indique la fraction que le taux urinaire représente par rapport au taux total d'élimination k. Finalement, les flèches indiquent le sens de circulation des molécules de la substance toxique.

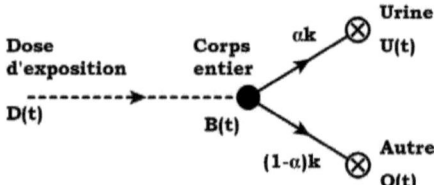

Figure 2-2: Modèle à compartiment unique.

Les unités utilisées pour simuler la quantité de xénobiotique dans les compartiments peuvent varier en fonction des mesures expérimentales disponibles. Ainsi, les équations correspondant à l'exemple donné par la figure 2-2 seraient :

$$\left(\frac{\partial}{\partial t} + k\right) B(t) = D(t), \tag{2-1}$$

38

$$\frac{\partial}{\partial t} U(t) = \alpha k B(t).$$ (2-2)

Dans les équations précédentes, si B(t) est en unités de moles et le temps est mesuré en heures (h), alors k est exprimé en h^{-1}, D(t) en mol h^{-1}, U(t) en mol et α est sans dimension. Si l'on considère l'exposition à une dose unique D_0 (mol) sur une période de temps infinitésimale, la solution aux équations 2-1 et 2-2 est simplement une élimination exponentielle (Garrett, 1994) caractérisée par la une demi-vie $t_{1/2}$ (Balant et Gexfabry, 1990; Norman, 1992) définie par le taux d'élimination k : $t_{1/2}=k^{-1}\ln(2)$.

$$B(t) = D_0 e^{-kt},$$ (2-3)

$$U(t) = \alpha D_0 (1 - e^{-kt}),$$ (2-4)

Avec les deux solutions précédentes (2-3 et 2-4), nous avons toute l'information nécessaire sur la quantité de la substance toxique dans l'organisme en tout temps, et la quantité excrétée par voie urinaire en tout temps (White, 1988).

En général, les compartiments représentent la quantité du composé toxique au niveau de différents sites à l'aide des fonctions analytiques. Ces deux fonctions obtenues sont caractérisées par deux paramètres inconnus et propres au xénobiotique en question : k et α. Des données expérimentales sont nécessaires à la détermination de ces paramètres.

2.6.1.2 Taux de transfert

Si le nombre de compartiments et le nombre de liens entre eux augmentent, les modèles à compartiment se complexifient. Normalement, le taux de transfert entre compartiments peut être associé à une ou plusieurs fonctions physiologiques distinctes (Rowland et Tozer, 2011; Tozer et Rowland, 2006). Usuellement, le taux de transfert représente l'ensemble des processus pouvant permettre à une substance de passer d'un compartiment à un autre. Ainsi, dans le cas présenté à la figure 2-2, le taux d'élimination urinaire αk devrait être proportionnel à l'élimination physiologique de l'urine, mais il ne représente pas uniquement cela. Le taux αk pourrait inclure la métabolisation, la liaison avec des protéines, l'irrigation sanguine des reins, le taux de filtration glomérulaire du composant toxique, la réabsorption tubulaire passive ou active, *etc.*

2.6.1.2.1 Absorption

Un bon exemple de l'utilisation des taux de transfert est la modélisation des voies d'entrée. Dans la figure 2-3, nous avons modifié le modèle à un seul compartiment en ajoutant un compartiment pour simuler l'absorption par voie orale (Rubinow, 2002). Dans cet exemple à deux compartiments, nous avons également laissé de côté l'organisme au complet, car nous voulons nous concentrer sur les quantités de xénobiotique observées dans le sang après son administration orale. Dans la figure 2-3, nous observons le nouveau compartiment qui représente tout le tractus gastro-intestinal au complet (de la bouche aux intestins). Ainsi, lors de l'absorption orale, le compartiment représentant le tractus GI(t) peut transférer le xénobiotique au compartiment du sang B(t) avec un taux d'absorption $\alpha'k'$ ou bien l'éliminer via les fèces F(t). Le paramètre α' est défini comme étant la proportion de k', ce dernier étant défini comme le taux de transfert de la lumière du tractus gastro intestinal vers la circulation sanguine.

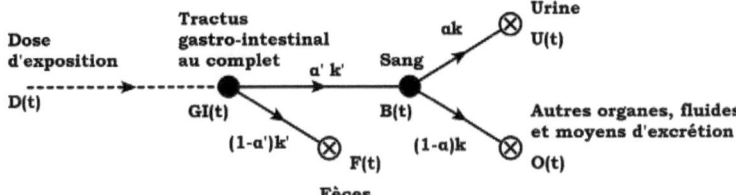

Figure 2-3: Modèle à deux compartiments.

Le taux d'absorption α'k' représente l'ensemble des processus de transport d'une substance toxique lors de son absorption orale. En d'autres termes, la valeur de α'k' symbolise en même temps : la durée de transit par le tractus gastro-intestinal, l'absorption passive et active par l'œsophage, l'estomac, le duodénum, l'intestin grêle, le gros intestin et l'éventuelle biotransformation dans le tractus. Ainsi, la seule information que ce paramètre nous permet de connaître est la vitesse d'absorption de la substance toxique sans nous indiquer le processus ou le site exact d'absorption.

La même observation s'applique aux autres voies d'absorption lors de la modélisation toxicocinétique à compartiments. Par exemple, lors d'exposition cutanée, si nous considérons un seul compartiment pour simuler l'absorption cutanée, le taux d'absorption inclurait la biotransformation dans la peau, le transfert actif ou passif, le temps de passage par l'épiderme, le derme et l'hypoderme. Si nous recherchons à connaitre les quantités du xénobiotique dans chaque partie de la peau ainsi que leur taux de transfert correspondant, nous devons inclure un nouveau compartiment par section dermique. Dans le cas présenté à la figure 2-3, les équations cinétiques du modèle à deux compartiments seraient (Laskarzewski *et al.*, 1982):

$$\left(\frac{\partial}{\partial t} + k'\right) GI(t) = D(t), \tag{2-5}$$

41

$$\left(\frac{\partial}{\partial t} + k\right) B(t) = \alpha' k' GI(t),$$

(2-6)

$$\frac{\partial}{\partial t} U(t) = \alpha k B(t).$$

(2-7)

Tel que dans le cas précédent, le nombre de paramètres à déterminer augmente proportionnellement au nombre de compartiments représentés, d'où le souci de n'inclure que des compartiments nécessaires à la modélisation. Les solutions de ces équations différentielles sont presque aussi simples que les fonctions 2-3 et 2-4 (voir, par exemple, Shimmins *et al.* (1967)).

2.6.1.2.2 Biotransformation

Dans certains cas où la dose d'exposition n'est pas élevée par rapport à la constante de Michaelis-Menten, la métabolisation d'un agent toxique peut être également représentée par un taux de transfert linéaire, tel que décrit dans les sections précédentes. Cependant, il est connu que la biotransformation de la plupart des substances est conduite par l'action d'enzymes (Casarett *et al.*, 1996; Klaassen, 2013). Ce processus est connu par sa non-linéarité et suit typiquement un comportement décrit par l'équation de Michaelis-Menten (Goulding, 1986; Hayes, 2008).

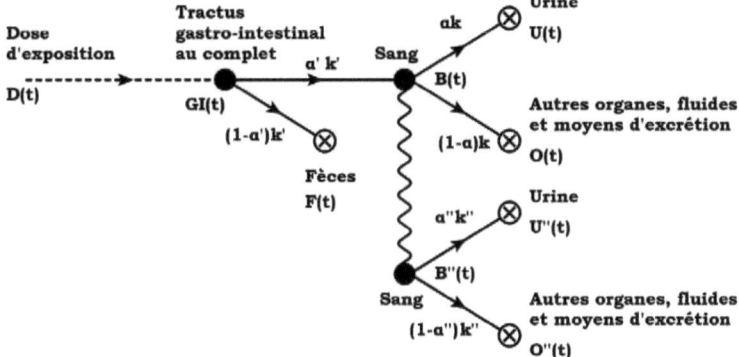

Figure 2-4: Modèle à trois compartiments.

Si nous voulons inclure explicitement une biotransformation non-linéaire, nous devons ajouter un nouveau compartiment dans le modèle à deux compartiments de la figure 2-3. Ainsi, la figure 2-4 montre le nouveau compartiment qui représente la quantité du métabolite de la substance toxique dans le sang B''(t). La ligne à trait ondulé représente la biotransformation de la substance toxique en son métabolite. Maintenant, nous avons deux manières d'éliminer la substance toxique par voie urinaire : élimination directe de la molécule mère U(t) ou élimination de son métabolite U''(t). Avec la configuration présentée à la figure 2-4, les équations différentielles du modèle sont les suivantes (Gerlowski et Jain, 1983):

$$\left(\frac{\partial}{\partial t} + k'\right) GI(t) = D(t), \tag{2-8}$$

$$\left(\frac{\partial}{\partial t} + k + \frac{V_{max}}{K_m + B(t)}\right) B(t) = \alpha' k' GI(t), \tag{2-9}$$

$$\left(\frac{\partial}{\partial t} + k''\right) B''(t) = \frac{V_{max}}{K_m + B(t)} B(t), \tag{2-10}$$

$$\frac{\partial}{\partial t} U(t) = \alpha k B(t).$$ (2-11)

$$\frac{\partial}{\partial t} U''(t) = \alpha'' k'' B''(t).$$ (2-12)

Dans les équations précédentes, tous les paramètres ont la même définition que dans le cas à deux compartiments incluant les paramètres doublement accentués qui correspondent au métabolite de l'agent toxique. Les paramètres associés à la métabolisation sont la vitesse maximale atteinte V_{max} par la métabolisation (en moles par heures, par exemple) et la constante de Michaelis-Menten K_m (en moles) qui représente le point auquel le taux est égal à la moitié de V_{max}.

Dans le cas des processus non linéaires de métabolisation, il est très difficile de résoudre les équations différentielles du modèle de manière analytique. Ainsi, un modèle aussi simple que celui décrit par la figure 2-4 doit être résolu par des méthodes numériques et les solutions aux équations différentielles ne peuvent plus nous révéler les relations entre les paramètres de manière explicite comme dans le cas linéaire du modèle à deux compartiments, à moins que l'on réussisse à linéariser le terme non-linéaire des équations 2-9 et 2-10 (Arfken *et al.*, 2012). Cette linéarisation est possible, par exemple, lorsque les quantités disponibles pour la biotransformation de la substance sont petites :

$$\frac{V_{max}}{K_m+B(t)} B(t) \bigg|_{B(t) \ll K_m} = \frac{V_{max}}{K_m} B(t) - \frac{V_{max}}{K_m^2} B^2(t) + \frac{V_{max}}{K_m^3} B^3(t) + O[B(t)]^4.$$ (2-13)

2.6.1.2.3 Excrétion

44

Dans la modélisation toxicocinétique à compartiments, la modélisation du taux de transfert qui représente l'élimination est la partie la plus importante du modèle puisqu'elle est déterminée par les mesures expérimentales accessibles. Ainsi, dans le modèle, il est nécessaire de définir clairement les paramètres représentés par le taux d'élimination utilisés lors de la modélisation et de les comparer à ceux définis par les mesures expérimentales déjà observées (Wells, 2012).

Par exemple, dans le modèle à un compartiment, le taux d'élimination urinaire αk a été défini en supposant que les collectes urinaires complètes du xénobiotique sont disponibles pour les comparer aux prédictions du modèle (figure 2-2). Cependant, si seulement les données provenant des fèces sont disponibles expérimentalement, αk devrait plutôt représenter le taux d'excrétion fécale du xénobiotique. Encore une fois, le même paramètre αk du modèle peut représenter des phénomènes complètement différents en fonction de l'interprétation que l'on fait de celui-ci. Ainsi, dans le cas de l'excrétion fécale, ce paramètre représenterait le temps de transit dans les intestins jusqu'à la sortie des fèces. Finalement, une fois que la substance toxique est collectée par la voie d'excrétion pertinente, rappelons que la dégradation de la substance et de ses métabolites peut affecter la concordance entre le taux d'élimination observé expérimentalement et celui modélisé.

2.6.1.3 Autres paramètres

Dans la modélisation toxicologique à compartiments, outre les taux de transfert et les compartiments, il y a d'autres paramètres dont il faut tenir compte lors de la modélisation (Godfrey, 1983). La dégradation de la substance toxique « en dehors de l'organisme », lors de son analyse chimique, pourrait être modélisée tout simplement par une constante représentant la « perte de la substance ». Au niveau de l'absorption de la substance, il peut

45

y avoir également des « pertes » du xénobiotique qui peuvent être modélisées par des fractions d'absorption correspondantes. Dans le cas de l'exposition cutanée, seulement une portion de la substance toxique peut être en contact direct avec la peau et être donc disponible à être absorbée. Ainsi, la dose administrée ne correspond pas avec la dose « réellement » disponible. Alors, ce type de phénomène peut être modélisé par des fractions d'absorption (Bouchard *et al.*, 2005). D'autres paramètres à considérer, lors de la modélisation à compartiments, comprennent des constantes qui représentent des fractions d'absorption (par exemple, lors de l'inhalation, seulement une partie de l'air entrant est en contact avec les sites d'absorption pulmonaire), des ratios entre métabolites (dans le cas de la modélisation de plusieurs métabolites à la fois), des mélanges de substances, *etc.* Dans les cas présentés précédemment, les facteurs α tenaient compte de la proportion que le taux urinaire représentait par rapport au taux total d'élimination, incluant les autres sites d'excrétion.

2.6.2 Modèles pharmacocinétiques à base physiologique

En 1937, le père de la pharmacocinétique, Torsten Teorell, introduit plusieurs idées qui rendaient des modèles à compartiments anciens plus *physiologiques* (Rescigno, 2010). Par exemple, il a représenté certaines substances dans des organes spécifiques par des compartiments séparés ; il a avancé l'idée de la transformation chimique du produit d'intérêt en d'autres molécules à une localisation donnée; il a considéré l'absorption sanguine d'une substance lors d'une administration sous-cutanée; il a modélisé un échange entre les tissues et le sang et une élimination sanguine des substances vers l'urine (Paalzow et Teorell, 1995). Malheureusement, les outils informatiques nécessaires pour résoudre les équations différentielles que Teorell avait établies n'existaient pas encore (Wagner, 1981). Ce fut seulement vers les années soixante-dix que les chimistes commencèrent à utiliser ces modèles pour prédire la distribution de plusieurs médicaments dans l'organisme (Andersen,

1995). L'utilisation de modèles physiologiques en évaluation des risques toxicologiques commence dans les années quatre-vingt avec la modélisation du chlorure de méthylène (Andersen *et al.*, 1987).

2.6.2.1 Représentation d'organes et de fluides corporels

Les modèles PCBP sont essentiellement des modèles à compartiments comme ceux décrits précédemment. Cependant, la structure des modèles PCBP est composée en fonction de la physiologie du sujet à modéliser. Ainsi, tout taux de transfert, tout compartiment et tout autre paramètre utilisé doit avoir une interprétation physiologique réelle (Aiache, 1985; Balant et Gexfabry, 1990). Tel que présenté à la figure 2-5, le plus simple modèle PCBP que l'on puisse construire, pour tenir compte de toutes les voies d'administration possibles, doit comporter au moins sept compartiments bien définis par des organes et des fluides corporels séparés.

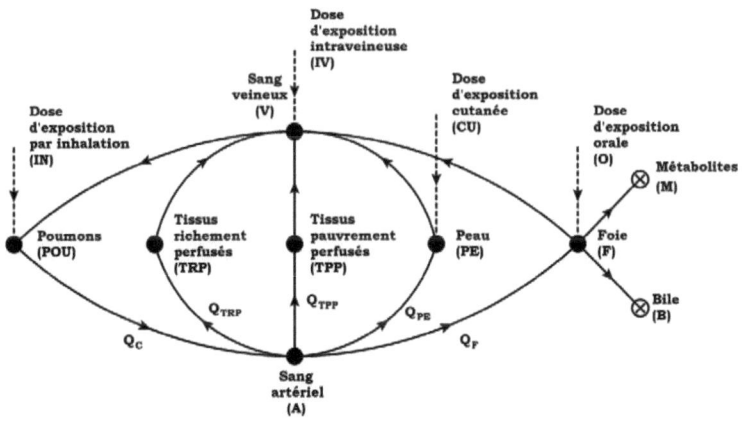

Figure 2-5: Modèle PCBP simple.

Typiquement, les compartiments des modèles PCBP représentent la concentration du xénobiotique dans les tissus et les fluides qui les définissent. Ainsi, dans le cas présenté à la figure 2-5, le compartiment POU représente la concentration de l'agent toxique aux poumons en fonction du temps $C_{POU}(t)$, $C_{TRP}(t)$ celle de tous les tissus richement perfusés (reins, intestins et cerveau, sans inclure le foie et les poumons), $C_{TPP}(t)$ celle des tissus pauvrement perfusés (les tissus adipeux, les muscles, le cœur, la rate, les os, sans inclure la peau), $C_{PE}(t)$ celle de la peau et $C_F(t)$ celle du foie. Généralement, les tissus ayant le même comportement cinétique et dynamique peuvent être combinés, comme dans le cas de tissus rapidement perfusés et ceux lentement perfusés. Dans le cas où la distribution du xénobiotique n'est pas restreinte par la diffusion (Albanese *et al.*, 2002; Gerlowski et Jain, 1983), les équations différentielles décrivant la variation du contenu dans les différents organes seraient :

$$V_{POU} \frac{\partial}{\partial t} C_{POU}(t) = Q_C \left[C_V(t) - \frac{C_{POU}(t)}{P_{POU}} \right] + Q_P \left[C_{IN}(t) - \frac{C_{POU}(t)}{P_P} \right], \quad (2\text{-}14)$$

$$V_{TRP} \frac{\partial}{\partial t} C_{TRP}(t) = Q_{TRP} \left[C_A(t) - \frac{C_{TRP}(t)}{P_{TRP}} \right], \quad (2\text{-}15)$$

$$V_{TPP} \frac{\partial}{\partial t} C_{TPP}(t) = Q_{TPP} \left[C_A(t) - \frac{C_{TRP}(t)}{P_{TPP}} \right], \quad (2\text{-}16)$$

$$V_{PE} \frac{\partial}{\partial t} C_{PE}(t) = Q_{PE} \left[C_A(t) - \frac{C_{PE}(t)}{P_{PE}} \right] + S\, K_P \left[C_{CU}(t) - \frac{C_{PE}(t)}{P_S} \right], \quad (2\text{-}17)$$

$$V_F \frac{\partial}{\partial t} C_F(t) = Q_F \left[C_A(t) - \frac{C_F(t)}{P_F} \right] - K_B C_F(t) - \frac{V_{max}}{K_m + C_F(t)} C_F(t) + K_O C_O(t), \quad (2\text{-}18)$$

$$V_A \frac{\partial}{\partial t} C_A(t) = Q_C [C_{POU}(t) - C_A(t)], \quad (2\text{-}19)$$

$$V_V \frac{\partial}{\partial t} C_V(t) = Q_{POU} \frac{C_{POU}(t)}{P_{POU}} + Q_{TRP} \frac{C_{TRP}(t)}{P_{TRP}} + Q_{TPP} \frac{C_{TPP}(t)}{P_{TPP}} + Q_{PE} \frac{C_{PE}(t)}{P_{PE}} + Q_F \frac{C_F(t)}{P_F} - Q_C C_V(t) + C_{IV}(t). \quad (2\text{-}20)$$

Dans les équations précédentes, les indices des variables indiquent le compartiment correspondant tel que décrit à la figure 2-5. Ainsi, V_{POU} est le volume des poumons, Q_F est le flux sanguin qui irrigue le foie, P_{PE} est le coefficient de partition associé à la peau. Les autres paramètres correspondent aux voies d'entrée et aux voies d'élimination seront décrits dans les sections suivantes. Le coefficient de partition représente le ratio entre la concentration du xénobiotique dans l'organe et la concentration du xénobiotique dans le sang veineux du tissu. Dans le cas où un ou plusieurs organes limitent la distribution de la substance toxique par la diffusion passive ou active à travers la matrice cellulaire, les équations différentielles doivent être modifiées pour accommoder ce phénomène (Espie *et al.*, 2009). Tel qu'illustré à la figure 2-6, il est nécessaire de diviser le compartiment en question en un compartiment d'échange avec le sang (E) et un compartiment décrivant la matrice cellulaire et limité par la diffusion (LD). Maintenant, en plus du coefficient de partition associé au compartiment limité par la diffusion P, il est nécessaire d'inclure le taux de transfert (coefficient de perméabilité) entre les deux compartiments K_{PA}.

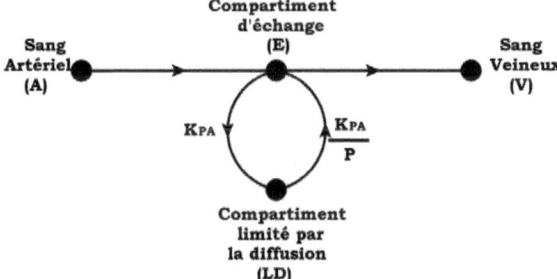

Figure 2-6: Compartiment limité par la diffusion.

Les équations différentielles qui représentent les compartiments limités par la diffusion sont (Albanese *et al.*, 2002; Thompson et Beard, 2011):

$$V_E \frac{\partial}{\partial t} C_E(t) = Q_E[C_A(t) - C_E(t)] - K_{PA}\left[C_E(t) - \frac{C_{LD}(t)}{P}\right], \qquad (2\text{-}21)$$

$$V_{LD} \frac{\partial}{\partial t} C_{LD}(t) = K_{PA}\left[C_E(t) - \frac{C_{LD}(t)}{P}\right], \qquad (2\text{-}22)$$

Finalement, il faut mentionner que le choix des tissus modélisés dépend des organes cibles attribués à chaque substance toxique ou des processus cinétiques d'intérêt. À tout moment, il est possible de grouper ou de diviser les compartiments du modèle PCBP pour mieux simuler la cinétique d'une région particulière à l'organisme (Reddy, 2005; Rescigno, 2010).

2.6.2.2 Absorption

Dans la modélisation PCBP, chaque voie d'exposition est modélisée de façon différente (Hayes, 2008; Meibohm et Derendorf, 1997; Reddy, 2005). L'exposition par voie cutanée nécessite forcément un échange entre le produit qui peut être absorbé et celui qui reste en surface (voir équation 2-17). Tel que montré à la figure 2-7, un transfert du xénobiotique de la surface de la peau au sang peut être modélisé grâce au coefficient de partition P_{PE} et entre la peau et le véhicule de la substance toxique grâce au coefficient de partition P_S. La perméabilité de la peau détermine la vitesse de pénétration du composé toxique dans la peau (K_P, normalement en cm/h) tandis que S détermine la surface de la peau en contact avec la substance toxique (S, normalement en cm^2).

50

Ainsi, si l'on élargit la surface de la peau exposée ou sa perméabilité, on augmente proportionnellement l'absorption cutanée. De même, si l'on diminue le coefficient de partition avec le sang, l'absorption du xénobiotique se verra favorisée. Au contraire, si l'on diminue le coefficient de partition avec le véhicule du xénobiotique, l'absorption se verra réduite.

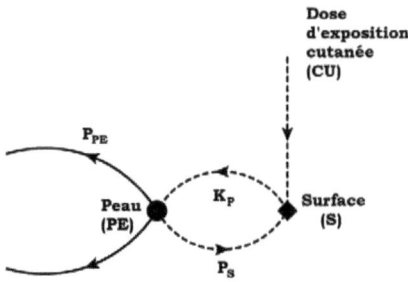

Figure 2-7: Absorption cutanée.

De la même façon, l'exposition par inhalation requiert un échange puisque la substance toxique peut-être absorbée et expulsée par les poumons (voir figure 2-8 et équation 2-14). Ainsi, le paramètre qui régule la respiration est le taux de ventilation pulmonaire (Q_P en unités de L/h). De ce qui est disponible du composant toxique dans les poumons, une partie peut être transférée au sang avec un taux régit par le coefficient de partition avec le sang P_{POU}. Une autre partie du xénobiotique peut être expulsée avec un taux régulé par le coefficient de partition P_P (Peters, 2011).

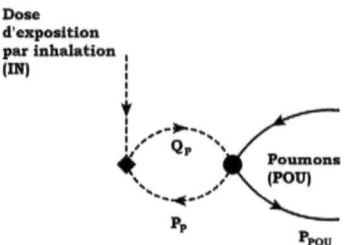

Figure 2-8: Absorption par inhalation.

Lors de l'exposition par ingestion, si le composé toxique est absorbé au complet par le tractus gastro-intestinal, la simulation de cette voie peut être représentée par un taux de transfert constant K_O, tel que décrit à l'équation 2-18. Sinon, si l'absorption est partielle, il est préférable d'inclure le compartiment du tractus gastro-intestinal pour permettre l'élimination du xénobiotique par les fèces et l'absorption directe par la veine porte (voir figure 2-9).

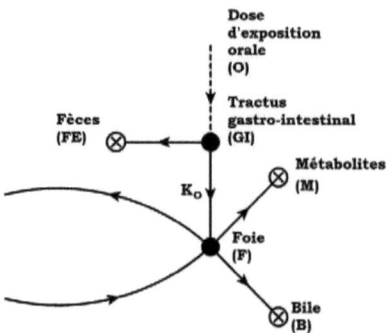

Figure 2-9: Absorption orale partielle.

52

D'autres voies d'entrée adéquates pour décrire la cinétique de plusieurs substances peuvent être simulées par les modèles PCBP. Cependant, les trois voies que nous venons de décrire sont les voies les plus pertinentes pour la modélisation toxicocinétique chez les êtres humains (Hayes, 2008).

2.6.2.3 Distribution

En général, la distribution de substances toxiques à tous les organes se fait via le flux sanguin. Il existe certains cas où le système lymphatique peut participer à la distribution de substances (tel est le cas de certaines molécules, comme le diazépam, de petit poids moléculaire (Lamka *et al.*, 1991)). Le système de distribution par le sang est cyclique puisqu'il permet le transit de molécules à répétition dans chaque organe (jusqu'au moment où il y a interaction). Essentiellement, chaque organe reçoit le sang artériel en provenance des poumons pour ensuite le retourner aux poumons par le sang veineux directement ou en passant d'abord par d'autres tissus avant. Ainsi, la circulation sanguine suit un sens bien défini. En fonction de la voie d'exposition, les substances toxiques peuvent rencontrer des organes différents lors de leur premier passage dans la circulation (Timbrell, 2002). Pour certains composants, ce phénomène de premier passage peut entraîner des concentrations importantes de xénobiotique dans les premiers organes rencontrés et correspondant à la voie d'exposition employée (par exemple, de fortes concentrations du BaP peuvent s'observer dans les poumons suite à une exposition intraveineuse (Moir *et al.*, 1998; Marie *et al.*, 2010)). Tout cela est implicitement inclus dans la modélisation PCBP puisque les équations utilisées représentent cet ordre physiologique dans l'acheminement de la substance toxique à travers la circulation sanguine (IPCS, 2010).

2.6.2.4 Biotransformation

La modélisation PCBP permet non seulement de modéliser la métabolisation au foie (principal site de détoxification de l'organisme), mais aussi aux autres sites présentant une activité enzymatique importante pour l'agent toxique (Klaassen, 2001). Tel est le cas du poumon, de la peau et du tractus gastro-intestinal (Lipscomb et Ohanian, 2007; Peters, 2011). La majorité des modèles PCBP représente le processus de métabolisation dans le compartiment hépatique par le mécanisme de saturation de Michaelis-Menten (Lipscomb et Ohanian, 2007; Reddy, 2005). Cette forme de modélisation est intrinsèquement non-linéaire. Cependant, en fonction des doses d'exposition que le modèle souhaite simuler, tel que décrit à la section 2.5.1.2.2, ce mécanisme peut également être modélisé par des taux de biotransformation de premier ordre (Notari, 1987).

La modélisation PCBP permet de simuler la production de plusieurs métabolites et de tracer leur devenir dans l'organisme. Par exemple, dans le cas du BaP, on pourrait simuler la cinétique de plusieurs métabolites ($i=1..n$) à la fois, en supposant qu'il n'existe pas de compétition entre les divers métabolites, via la définition d'une constante de Michaelis-Menten $K_m^{(i)}$ et d'une vitesse maximale de production $V_{max}^{(i)}$ pour chacun :

$$V_F \frac{\partial}{\partial t} C_F(t) = Q_F \left[C_A(t) - \frac{C_F(t)}{P_F} \right] - K_B C_F(t) - \sum_{i=1}^{n} \frac{V_{max}^{(i)}}{K_m^{(i)}+C_F(t)} C_F(t) + K_O C_O(t), (2\text{-}23)$$

Lors d'une exposition réelle, les substances toxiques sont généralement absorbées en mélanges ou bien elles sont absorbées dans l'organisme contenant déjà d'autres substances pouvant interférer entre elles. Les modèles PCBP nous permettent de clairement identifier chaque substance ainsi que leurs métabolites séparément. Il est possible de modéliser

54

également des phénomènes d'induction ou d'inhibition enzymatique (Aiache, 1985; Krishnan *et al.*, 1994; Reddy, 2005).

2.6.2.5 Excrétion

Dans la modélisation PCBP, l'élimination des substances toxiques est exprimée par des taux constants d'excrétion communément associés à la clairance (Notari, 1987). La clairance est un concept fort utile dans la cinétique d'extraction des xénobiotiques par un organe spécifique (principalement les reins et le foie). La clairance est définie par le volume de liquide épuré du xénobiotique, dans un temps donné (Hayes, 2008). Dans la plupart des cas, nous parlons de la clairance rénale ou hépatique et le liquide à épurer est directement le sang. La clairance Cl est souvent associée au concept de volume de distribution apparent V_d qui est défini par le volume fictif dans lequel serait dissoute la quantité administrée de substance toxique pour donner la concentration observée dans le sang.

$$Cl = k_{elim} \, V_d, \tag{2-24}$$

Ainsi, le taux d'élimination k_{elim}, pour une voie donnée est défini par la clairance divisée par le volume apparent de distribution.

Pour l'élimination rénale, la clairance utilisée dans les modèles PCBP doit prendre en compte la filtration glomérulaire, la sécrétion tubulaire et la réabsorption tubulaire active et passive (Notari, 1987). Dans le cas de l'élimination par le foie (en excluant le processus de métabolisation), il faut tenir compte de l'excrétion biliaire et parfois aussi du cycle entéro-

55

hépatique. D'autres sites d'excrétion existent, mais sont soit de moindre importance étant donné les quantités limitées de substances qu'ils éliminent (excrétion salivaire, mammaire, sudorifique), soit dépendent directement de la voie d'absorption (les poumons, la peau, le tractus gastro-intestinal). Il est habituel de modéliser ces processus en même temps que ceux de l'absorption (Aiache, 1985).

2.6.3 Modèles actuellement utilisés pour la modélisation cinétique du BaP

Récemment, des avancées ont été faites dans la modélisation du BaP. D'une part, la modélisation du BaP a été réalisée dans l'environnement : le sol, (Landrum, 1989; Lu *et al.*, 2004), l'air (Hauck *et al.*, 2008; San Jose *et al.*, 2013) et l'eau (Ciffroy *et al.*, 2011; Gerde et Scholander, 1988; Heinonen *et al.*, 2000). D'autre part, des modèles ont tenté d'approfondir les connaissances sur la toxicocinétique BaP. Dans les sections suivantes, nous décrirons les grandes lignes des deux types de modélisation utilisées.

2.6.3.1 Modèles à compartiments

Avec l'apparition des modèles PCBP, l'utilisation des modèles toxicocinétiques à compartiments classiques a diminué. Cependant, ils sont encore très utilisés pour obtenir des paramètres toxicocinétiques reliés aux profils sanguins incluant certains paramètres cinétiques tels que la demi-vie d'élimination, l'aire sous la courbe, le temps de séjour moyen (Reisfeld et Mayeno, 2012). Ainsi, la majorité des modèles toxicocinétiques à compartiments du BaP visent à obtenir les paramètres d'élimination du BaP à partir de l'ajustement à des fonctions exponentielles.

56

Tableau 2-12 : Modèles à compartiments du BaP.

Espèce	Voie d'exposition	Dose d'exposition (BaP)	Modèle	Demi-vie d'élimination (h)	Référence
Rat Fischer 344	Orale	100 mg/kg	Méthodes non compartimenta les	5,8 à partir du plasma	Ramesh *et al.* (2001b)
Rat Sprague Dawley	IV	40 μmol/kg	Deux compartiments	12,5 à partir du sang	Marie *et al.* (2010)
Rat Wistar	IV	2, 6 et 15 mg/kg marqué au ^{14}C	Deux compartiments	10,2, 15,2 et 11,8, respectivement, à partir du sang	Moir *et al.* (1998)
Rat Sprague Dawley	IV	2 mg/kg	Méthodes non compartimenta les	3,3 à partir du sérum	Cao *et al.* (2005)
Rat Fischer 344	Inhalation	0,1, 1 et 2,5 mg/m^3	Méthodes non compartimenta les	0,6 à partir du plasma	Ramesh *et al.* (2001a)
Rat Sprague Dawley	Instillation intra-trachéale	1 μg/kg marqué au ^3H	Trois compartiments	4,6	Weyand et Bevan (1986)
Rat Sprague Dawley	Injection	117 nmol/kg marqué au ^3H	Deux compartiments	1,8-3,3	Wiersma et Roth (1983)
Rat Sprague Dawley	IV	0,01 et 0,05 mg/kg marqué au ^{14}C	Un compartiment	7,7 et 9,2 respectivement pour le 3-OHBaP	Payan *et al.* (2009)

Grâce à l'utilisation d'un modèle à deux compartiments (tel que celui décrit aux équations 2-5, 2-6 et 2-7), Wiersma et Roth (1983) ont calculé une demi-vie entre 1,8 et 3,3 heures, à partir des données sanguines chez des rats exposés par injection au BaP marqué au ^3H. Des valeurs plus élevées ont été obtenues par Marie *et al.* (2010) et Moir *et al.* (1998). Dans la première étude, ils ont obtenu une demi-vie de 12,5 heures, à partir de l'ajustement à la somme de deux fonctions exponentielles de profils sanguins de rats exposés par voie intraveineuse. Dans la deuxième étude, le même ajustement a donné des valeurs de demi-vie oscillant entre 10 et 15 heures, pour des doses d'exposition différentes. Lors d'une exposition intra-trachéale au BaP chez des rats, Weyand et Bevan (1986) ont obtenu une demi-vie d'élimination de 4,6 heures, suite à un ajustement d'un modèle à trois compartiments. En fait, le modèle consistait en la somme de trois fonctions exponentielles. La plus petite des trois constantes dérivées de ces décroissances exponentielles a été identifiée comme étant la demi-vie d'élimination.

Cao *et al.* (2005) ont calculé une demi-vie de 3,3 heures, à partir du sérum de rats exposés par voie intraveineuse au BaP, par des méthodes non compartimentales. Techniquement, par ces méthodes, la demi-vie peut être calculée par la formule suivante :

$$t_{\frac{1}{2}} = \frac{1}{k_{elim}} = \frac{V_d}{Cl} = \frac{AUC\,V_d}{D_0}, \tag{2-25}$$

Dans l'équation précédente, la clairance peut être calculée par le ratio entre la dose d'exposition D_0 et l'aire sous la courbe du profil sanguin AUC. Dans une autre analyse par techniques non compartimentales, Ramesh *et al.* (2001a) ont obtenu une demi-vie d'élimination plasmatique de 0,6 heures dans une étude chez des rats exposés par inhalation. Les mêmes auteurs et d'autres collaborateurs ont analysé les profils sanguins des rats exposés par voie orale au BaP (Ramesh *et al.*, 2001b); ils ont calculé une demi-vie d'élimination plasmatique de 5,8 heures, par régression linéaire sur les profils sanguins. Dans une étude par exposition intraveineuse de rats, Payan *et al.* (2009) ont calculé une demi-vie d'élimination pour le 3-OHBaP de 7,7 et 9,2 heures, pour deux doses différentes du BaP marqué au ^{14}C (0,01 et 0,05 mg/kg).

2.6.3.2 Modèles PCBP

La première allusion à un modèle PCBP pour le BaP, dans les publications scientifiques, date de 1990 dans une étude sur l'activité métabolique des poumons et du foie (Roth et Vinegar, 1990). Ces chercheurs ont modélisé le profil temporel du BaP dans les poumons et le foie étant donné la contribution de ces organes à la biotransformation du BaP. Ils ont incorporé à leurs simulations un compartiment pour les tissus adipeux, étant donné leur capacité à cumuler certaines substances toxiques. Ils ont également inclus la possibilité d'une liaison du BaP au composant du sang artériel, du sang veineux, du foie et des poumons. Les valeurs de paramètres physiologiques ont été obtenues à partir d'expériences *in vitro* (Wiersma et Roth, 1983). Les chercheurs ont rapporté que les simulations présentées dans cette étude étaient adéquates pour les deux voies d'exposition analysées (intraveineuse et intra-artérielle) sauf pour les poumons. Ils ont aussi utilisé leur modèle pour simuler les profils temporels du BaP dans le sang et les tissus observés dans une autre étude chez le rat (celle de Schlede *et al.*, 1970). Cependant, le modèle n'a pas pu reproduire ces derniers profils, sans l'ajustement additionnel des paramètres obtenus à partir des expériences *in vitro* de Wiersma et Roth (1983). Malheureusement, ces auteurs n'ont pas présenté les valeurs de tous les paramètres employés dans leurs simulations. Il est donc impossible de reproduire leur modèle PCBP.

Plus récemment, Pery *et al.* (2011) ont utilisé la même structure de modèle PCBP proposée par Roth et Vinegar pour évaluer les effets du BaP sur l'expression des gènes codant pour IL-1β et NCF1. Par conséquent, ils ont modélisé les mêmes organes et fluides et ont utilisé les données cinétiques obtenues lors de l'exposition de rats au BaP, par instillation intra-trachéale (Weyand et Bevan, 1986), en présumant que la voie principale d'exposition au BaP est l'inhalation. Le modèle PCBP employé compte un seul compartiment limité par la diffusion correspondant aux tissus adipeux. Les valeurs physiologiques du modèle ont été obtenues des diverses publications scientifiques tandis que les coefficients de partition et le taux de biotransformation ont été calculés en utilisant la méthode du maximum de vraisemblance. Leur modélisation a seulement considéré une voie d'élimination : les intestins. Les simulations effectuées par le modèle PCBP ont bien représenté les profils temporels du BaP dans les tissus analysés (foie, poumons et sang). Cependant, la valeur de métabolisation considérée dans le modèle a dû être modifiée (cinq fois plus petite) pour pouvoir représenter un ensemble de données indépendant (soit les profils temporels du BaP dans le foie, les poumons, le sang et les tissus adipeux de Schlede *et al,.* 1970).

Tableau 2-13 : Modèles PCBP du BaP.

Espèce	Voie d'exposition	Dose d'exposition (BaP)	Modèle	Paramètres	Référence
Pétoncle	Immergés dans l'eau de mer	50 ng/L	Les branchies, la glande digestive, les muscles adducteurs et l'hémolymphe.	Demi-vie d'élimination= 34,7 h.	Liu *et al.* (2014)
Rat Sprague Dawley	IV	10 µg	Les poumons, les tissus adipeux, le foie, les tissus richement perfusés.	Coefficient de partition du foie = 134, Vitesse maximale de métabolisme = 0,087 mg/min, Constante de Michaelis-Menten = 0,0014 mg/mL.	Pery *et al.* (2011)
Rat Sprague Dawley	IV	117 nmol/kg	Les poumons, les tissus adipeux, le foie, les tissus richement perfusés.	Aucun présenté.	Roth et Vinegar (1990)
Rat Sprague Dawley	IV et Orale	0,06 mg/kg	Les poumons, les tissus adipeux, le foie, les tissus richement vascularisés, les tissus pauvrement perfusés.	Coefficient de partition du foie = 13,3, Vitesse maximale de métabolisme = 44,7 nmol/(min mL), Constante de Michaelis-Menten = 5,5 nmol/mL.	Crowell *et al.* (2011)

La même année, Crowell *et al.* (2011) ont réalisé un modèle PCBP préliminaire basé sur une structure identique, mais en ajoutant la voie d'exposition orale chez des rats. Les chercheurs ont commencé par construire le modèle PCBP par voie intraveineuse pour ensuite incorporer la voie orale. Ils ont modélisé un compartiment pour les poumons, le foie, les tissus adipeux, les tissus richement perfusés et ceux pauvrement perfusés. La distribution du

BaP dans tous les tissus a été considérée comme limitée par la perfusion de chaque organe, sauf pour le cas des tissus adipeux, qui ont été modélisés comme étant limités par la diffusion (comme dans les modélisations PCBP précédentes) à cause de leur rôle de réservoir de produits hautement lipophiles. Les auteurs ont mis l'accent sur l'utilisation des données expérimentales pour la détermination des paramètres du modèle. Les paramètres physiologiques, tels que le volume de tissus et la fraction du flux sanguin, ont été obtenus à partir de bases de données communément utilisées pour la construction de modèles PCBP (Brown *et al.*, 1997; Davies et Morris, 1993). Les constantes de biotransformation ont été obtenues à partir des expériences *in vitro* tirées de l'étude de Wiersma et Roth (1983). Les autres paramètres concernant l'absorption et l'élimination du BaP ont été optimisés à partir de diverses publications (Foth *et al.*, 1988; Osinski *et al.*, 2002; Roth *et al.*, 1993) par observation et par la méthode du maximum de vraisemblance sur le logarithme des profils expérimentaux. Les chercheurs ont également modélisé la liaison du BaP aux lipoprotéines du sang. Les simulations obtenues ont globalement bien représenté les profils sanguins provenant de plusieurs études expérimentales (Moir *et al.*, 1998; Schlede *et al.*, 1970; Wiersma et Roth, 1983). Cependant, les simulations concernant le foie et les poumons ont sous-estimé les concentrations observées suite à une exposition intraveineuse ou orale.

3 Problématique et objectifs de recherche

3.1 Problématique de recherche

3.1.1 Problématique générale

La surveillance biologique est reconnue comme outil privilégié pour évaluer l'exposition aux HAP dans la population générale et chez les travailleurs. Le BaP peut être utilisé comme indicateur de l'exposition aux HAP cancérogènes. Or, les concentrations urinaires de ce biomarqueur d'exposition dépendent de la cinétique du BaP et du 3-OHBaP dans l'organisme. Par conséquent, il est fondamental d'établir une relation entre la quantité de BaP absorbé dans l'organisme et la concentration urinaire de son métabolite pour chaque voie d'exposition chez l'humain. À l'aide de la modélisation toxicocinétique, il devient possible de reconstituer les doses d'exposition au BaP chez les individus à partir de mesures de son métabolite urinaire, le 3-OHBaP, utilisé comme biomarqueur d'exposition. Actuellement, il n'existe pas de modèle cinétique qui permette de retracer la relation entre le BaP et ses métabolites, notamment le 3-OHBaP.

Les modèles PCBP actuels (voir section 2.6.3.2) n'ont pas simulé la cinétique des métabolites du BaP. Ces modèles ont été construits principalement à partir de données expérimentales basées sur des mesures *in vitro* ou par extrapolation à partir de la structure chimique du BaP. Dans le dernier cas, l'activité biologique (par exemple, la liaison du produit toxique avec un récepteur) ou la réactivité chimique du BaP est quantitativement corrélée à la structure chimique d'un produit lui ressemblant. Tel que le démontrent les résultats obtenus jusqu'à ce jour avec ces modèles PCBP, ce type de données expérimentales n'est pas nécessairement représentatif de la cinétique *in vivo* du BaP (Crowell *et al.*, 2011;

Pery *et al.*, 2011; Roth et Vinegar, 1990). Dès lors, le développement de modèles cinétiques plus simples se basant sur des données expérimentales *in vivo* (déjà existantes chez les animaux) est nécessaire.

Enfin, l'utilité des diverses approches de modélisation toxicocinétique pour reconstruire les doses d'exposition à partir de mesures de biomarqueurs d'exposition est encore peu démontrée. Par conséquent, il apparait important de comparer les modèles dits à compartiments et les modèles PCBP pour déterminer leur capacité à reconstruire les scénarios d'exposition humaine. En d'autres termes, il faudrait comparer les concentrations urinaires de 3-OHBaP simulés à l'aide de ces différents modèles en considérant les mêmes scénarios d'exposition au BaP.

3.1.2 Problématiques spécifiques

3.1.2.1 Absence des modèles cinétiques pour le métabolite 3-OHBaP

D'importants résultats ont été obtenus dans diverses études concernant le taux d'excrétion fécal du BaP (Cao *et al.*, 2005; James *et al.*, 1995; Marie *et al.*, 2010; Moir *et al.*, 1998; Ramesh *et al.*, 2001a; Ramesh *et al.*, 2002; Ramesh *et al.*, 2001b; Schlede *et al.*, 1970; Wiersma et Roth, 1983) et le taux d'excrétion urinaire du 3-OHBaP (Chien et Yeh, 2012; Lee *et al.*, 2003; Payan *et al.*, 2009). La plupart de ces études toxicocinétiques ont tenté de déterminer ces taux d'excrétion par l'ajustement des profils sanguins aux sommes de courbes exponentielles du type :

$$B(t) = \sum_{\forall n} C_n e^{-k_n t} \tag{3-1}$$

Où, B(t) représente le profil temporel de la concentration sanguine du BaP, k_n est la constante de décroissance exponentielle, associée à la demi-vie, et C_n est une constante arbitraire. D'autres modèles PCBP ont abordé plus en détails la cinétique du BaP et son élimination. Ces modèles ont aussi estimé la constante d'élimination du BaP. À présent, il n'existe aucun modèle cinétique pour établir la relation entre l'exposition au BaP et les concentrations urinaires du 3-OHBaP, quelle que soit la voie d'absorption. Ainsi, il serait nécessaire d'avoir un modèle cinétique représentant l'absorption du BaP par de multiples voies d'exposition, la métabolisation en 3-OHBaP et l'élimination sous forme de BaP et de 3-OHBaP.

3.1.2.2 Insuffisance des modèles PCBP pour le BaP et pour le 3-OHBaP

Il existe deux modèles PCBP pour simuler la cinétique du BaP (Crowell *et al.*, 2011; Pery *et al.*, 2011) mais aucun pour simuler celle du 3-OHBaP. La la cinétique du BaP par voie intraveineuse a servi au développement initial du modèle et donc du comportement de ce composé dans les organes internes, afin de se départir de l'impact de la voie d'entrée sur les profils temporels évalués. Par la suite, les auteurs ont modifié le modèle pour inclure une exposition orale au BaP. Le second est un modèle PCBP qui se concentre principalement sur l'inhalation du BaP, sans inclure d'autres voies possibles d'exposition. Ils sont tous deux construits à partir d'études *in vitro* réalisées sur des préparations tissulaires de rats. Ainsi, il serait essentiel de construire un modèle PCBP basé sur des données cinétiques expérimentales *in vivo*, et ce, pour toutes les voies possibles d'exposition : respiratoire, cutanée et orale. Ce modèle devrait considérer la biotransformation du BaP en 3-OHBaP, ainsi que la cinétique suivant cette métabolisation. Enfin, ce modèle PCBP, basé sur des expériences sur des animaux, devrait pouvoir être extrapolé pour simuler cette cinétique chez l'être humain.

3.1.2.3 L'utilisation des modèles cinétiques du BaP en dosimétrie inverse

La reconstruction de la dose d'exposition, chez les travailleurs et chez la population en général, est une application maintenant reconnue des modèles toxicocinétiques en surveillance biologique (Bouchard *et al.*, 2006; Cote *et al.*, 2014; McNally *et al.*, 2012; McNally *et al.*, 2011; Molokanov *et al.*, 2010). Afin de déterminer la dose d'exposition chez des individus, à partir de mesures de concentrations des biomarqueurs dans le sang ou dans l'urine (dosimétrie inverse), il est nécessaire d'avoir un modèle à compartiments simple ou un modèle PCBP capable de décrire la cinétique du produit toxique *in vivo* chez l'humain. Dans le cas du BaP, il n'existe pas de modèle cinétique pouvant servir à cette fin. Ainsi, la construction de modèles cinétiques permettant de simuler les profils temporels de métabolites urinaires chez l'humain serait une avancée importante car elle permettrait prédire des scénarios d'exposition plausibles chez les individus exposés au BaP.

3.2 Hypothèse de recherche

L'hypothèse principale de cet ouvrage peut être émise comme suit : le développement de modèles toxicocinétiques permet de mieux comprendre les déterminants biologiques essentiels de la cinétique du 3-OHBaP comme biomarqueur d'exposition et donc de faciliter l'interprétation de données de surveillance biologique chez les individus exposés.

3.3 Objectifs de recherche

3.3.1 Objectif général

L'objectif cette thèse est de développer, comparer et appliquer deux outils de modélisation cinétique basés sur des expériences *in vivo*, à savoir la modélisation toxicocinétique à compartiments et la modélisation pharmacocinétique à base physiologique. Ces modèles doivent pouvoir évaluer les doses absorbées et associées à l'exposition professionnelle au BaP, à partir des mesures du biomarqueur 3-OHBaP.

3.3.2 Objectifs spécifiques

i. Développer un modèle toxicocinétique à plusieurs compartiments pour décrire la cinétique du BaP et de son métabolite le 3-OHBaP chez le rat en considérant différentes voies d'exposition. Pour ce faire, nous utiliserons des données *in vivo* de profils cinétiques du BaP et 3-OHBaP dans les poumons, les tissus adipeux, les reins, la peau, le foie, le sang, les fèces et l'urine, disponibles dans les publications scientifiques.

ii. Développer un modèle PCBP pour décrire la cinétique du BaP et de son métabolite 3-OHBaP chez le rat suite à des exposition par différentes voies, en utilisant les mêmes profils cinétiques *in vivo* publiés dans la littérature scientifique et en fonction des résultats obtenus avec le modèle toxicocinétique à plusieurs compartiments.

iii. Extrapoler à l'humain les modèles cinétiques établis à partir des profils cinétiques obtenus dans des expériences *in vivo* chez le rat.

66

iv. Déterminer l'importance de la voie d'exposition des travailleurs exposés au BaP, en fonction des tâches effectuées, en utilisant la modélisation cinétique et les concentrations urinaires de 3-OHBaP recueillies sur le terrain.

v. Comparer les profils urinaires du 3-OHBaP simulés à l'aide des deux modèles cinétiques développés.

3.4 Organisation du livre

Le présent ouvrage est formée selon un format *par articles*. Le corps du livre est composé de trois articles poursuivant les objectifs mentionnés précédemment et représentatifs du travail effectué durant les études doctorales.

Le premier article réfère à l'élaboration d'un modèle toxicocinétique à compartiments, pour le BaP et le 3-OHBaP à partir de données expérimentales *in vivo*. Il nous a permis d'identifier les caractéristiques fondamentales de la cinétique du BaP et du 3-OHBaP chez les rats. À partir de cette modélisation, nous avons pu élaborer un modèle décrivant la cinétique du BaP et du 3-OHBaP chez l'être humain par extrapolation du modèle animal. Ce modèle à compartiments a été utilisé pour reconstruire les doses absorbées par un employé exposé au BaP durant son travail. Nous avons constaté que les tissus adipeux et les poumons peuvent agir à titre de réservoirs pour le BaP non métabolisé, tandis que le reste du BaP est distribué dans tout l'organisme et biotransformé principalement par le foie. Enfin, nous avons pu constater que le 3-OHBaP produit suit une distribution rapide dans l'organisme mais également une importante élimination par la voie urinaire.

Dans le deuxième article, nous avons construit un modèle PCBP pour toutes les voies d'absorption (intraveineuse, respiratoire, cutanée et orale) à partir des mêmes profils cinétiques obtenus des expériences *in vivo* sur des rats. Grâce à ce modèle PCBP, nous avons pu vérifier la contribution majoritaire du foie dans la métabolisation du BaP par rapport à la métabolisation par d'autres organes. Enfin, nous avons exploré le rôle des reins dans l'élimination du 3-OHBaP par la voie urinaire et le phénomène de retardement observé dans l'excrétion. Nous avons également considéré l'élimination du 3-OHBaP à travers sa métabolisation par le foie.

Dans le troisième article, nous avons extrapolé un modèle PCBP et un modèle à compartiment unique représentant la cinétique du BaP et du 3-OHBaP chez le rat pour décrire la cinétique attendue chez l'humain. Nous avons ainsi pu reproduire les profils d'excrétion urinaire d'une dizaine d'individus exposés aux HAP au cours de leurs journées de travail. Enfin, nous avons déterminé les concentrations d'exposition de BaP nécessaires pour obtenir les profils urinaires du 3-OHBaP observés chez ces travailleurs. De plus, nous avons pu comparer les deux modèles cinétiques. Il en est ressorti qu'ils sont aussi adéquats pour reproduire les concentrations du BaP observées et les profils urinaires du 3-OHBaP collectés. Dans cet article, nous avons constaté que l'exposition cutanée indirecte du BaP est présente dans les usines étudiées.

Corps du livre

4 Premier article : Modeling of the internal kinetics of benzo(a)pyrene and 3-hydroxybenzo(a)pyrene biomarker from rat data

Heredia-Ortiz R., Bouchard M., Marie-Desvergne C., Viau C., Maître A.

Heredia-Ortiz R., Bouchard M., Marie-Desvergne C., Viau C., Maître A. Modeling of the internal kinetics of benzo(a)pyrene and 3-hydroxybenzo(a)pyrene biomarker from rat data. Toxicol Sci. 2011 Aug;122(2):275-87. doi: 10.1093/toxsci/kfr135. Epub 2011 May 24.

Modeling of the Internal Kinetics of Benzo(a)pyrene and 3-Hydroxybenzo(a)pyrene Biomarker from Rat Data

Roberto Heredia-Ortiz,[*] Michèle Bouchard,[*,1] Caroline Marie-Desvergne,[†] Claude Viau,[*] Anne Maître[†]

[*] Département de santé environnementale et santé au travail, Chaire d'analyse et de gestion des risques toxicologiques and Institut de recherche en santé publique de l'Université de Montréal (IRSPUM), Faculté de Médecine, Université de Montréal, C.P. 6128, Succursale Centre-ville, Montréal, Québec, Canada, H3C 3J7

[†] Equipe environnement et prédiction de la santé des populations, Laboratoire TIMC (UMR 5525), CHU de Grenoble, Université Joseph Fourier, Domaine de la Merci, 38700 La Tronche, France.

Short title: Kinetic model for BaP and 3-OHBaP biomarker

[1]Corresponding author:

Michèle Bouchard, Ph.D.

Associate professor

Department of Environmental and Occupational Health

University of Montreal

P.O Box 6128, Main Station, Montreal (QC)

H3C 3J7, CANADA

Telephone number: (514) 343-6111 ext 1640

Fax number: (514) 343-2200

4.1 Abstract

Measurements of 3-hydroxybenzo(a)pyrene (3-OHBaP) in urine has been proposed for the biomonitoring of exposure to benzo(a)pyrene (BaP) in workers. To allow a better understanding of the toxicokinetics of BaP and its key biomarker, a multi-compartment model was developed based on rat data previously obtained by this group. According to the model, intravenously injected BaP is rapidly distributed from blood to tissues ($t_{1/2}$ = 3.65 h), with particular affinity for tissue lipid components and liver and lung proteins. BaP is then rapidly distributed to lungs, where significant tissue uptake occurs, followed by the skin, liver and adipose tissues. Once in liver, BaP is readily metabolized, and 3-OHBaP is formed with a $t_{1/2}$ of 3.32 h. Lung metabolism of BaP was also accounted for, but its contribution to the whole kinetics was found to be negligible. Once formed, 3-OHBaP is distributed from blood to the various organs almost as fast as the parent compound ($t_{1/2}$ = 2.26 h). In kidneys, 3-OHBaP builds-up as a result of the smaller rate of 3-OHBaP urinary excretion ($t_{1/2}$ = 4.52 h) as compared to its transfer rate from blood to kidneys ($t_{1/2}$ = 27.8 min). However, overall clearance of 3-OHBaP from the body is driven by its biliary transfer from liver to the gastrointestinal tract ($t_{1/2}$ = 3.81 h). The model provides a great fit to an independent set of published data on 3-OHBaP urinary excretion time course (χ^2 = 0.019). This model proves useful in establishing the main biological determinants of the overall kinetics of these compounds.

Key Words: Benzo(a)pyrene; 3-hydroxybenzo(a)pyrene; toxicokinetic model; rat.

4.2 Introduction

Biomonitoring is recognized as a privileged approach to assess health risks arising from occupational exposure to polycyclic aromatic hydrocarbons (PAH) (Angerer et al., 2007; Jacob and Seidel, 2002). Measurements of 1-hydroxypyrene (1-OHP), a metabolite of the PAH pyrene, has been proposed for the biomonitoring of exposure to PAHs (Jongeneelen et al., 1987). Numerous biomonitoring studies in workers of different settings (aluminum plant, coke plant, wood treatment plant,…) have shown the usefulness of this biomarker to identify exposed groups of workers (Bouchard and Viau, 1999). However, given that pyrene is not a carcinogenic PAH, measurement of its metabolite 1-OHP for the biomonitoring of exposure to carcinogenic PAH has been questioned. Some authors have thus developed methods to identify other biomarkers of exposure to carcinogenic PAH (Bouchard et al., 2009; Gendre et al., 2002, 2004). This includes the development of sensitive methods allowing the measurement of urinary 3-hydroxybenzo(a)pyrene (3-OHBaP) as an indicator of occupational exposure to BaP (see Fig. 1 for chemical structures), one of the most studied and abundant carcinogenic PAH in workplaces (Forster et al., 2008; Gendre et al., 2002, 2004; Lafontaine et al., 2004; Maître et al., 2008). However, in the study of Gendre et al. (2004) in workers occupationally exposed to PAH, it was shown that 3-OHBaP urinary excretion was delayed compared to that of 1-OHP, with maximum excretion lagging by about 15 h that of 1-OHP.

To help interpret biomonitoring data on 3-OHBaP such as the one published by Gendre et al. (2004), the detailed time profiles of BaP and its biomarker 3-OHBaP have recently been documented in blood, tissues and excreta of intravenously exposed rats (Marie et al., 2010). The latter data complemented previously published urinary time course data of 3-OHBaP in rats intravenously injected with BaP under the same exposure conditions or with 3-OHBaP metabolite (Bouchard and Viau, 1996). The atypical urinary time course data previously reported by Bouchard and Viau (1996) following intravenous injection with an initial increase in urinary excretion rates in the hours post-dosing followed by a gradual elimination matched the kidney time course data recently documented by Marie et al. (2010).

Ramesh et al. (2001) also determined the time courses of the parent compound BaP and total aqueous and organic metabolites in blood, tissues and excreta of rats following a very high oral dose of BaP (400 µmol/kg), along with the relative distribution of specific BaP metabolites including 3-OHBaP; nonetheless, the specific time courses of the different metabolites were not reported. Other researchers have performed mass-balance studies following administration of labelled BaP (^{14}C-BaP or ^{3}H-BaP) but the kinetics of specific metabolites to be used as biomarkers were not quantified specifically (Mitchell, 1982; Moir et al., 1998; Uziel and Haglund, 1988; Withey et al., 1993).

While animal kinetic data are now available for 3-OHBaP (Bouchard and Viau 1996; Lee et al., 2003; Marie et al., 2010), the main biological processes governing the overall kinetics of 3-OHBaP in blood, tissues and excreta have not yet been fully elucidated. Knowledge of the kinetics of 3-OHBaP is needed for efficient use of 3-OHBaP as a potential biomarker of exposure to BaP. To better understand the biological determinants of the kinetic behavior of 3-OHBaP, the objective of this study was to develop a biomathemical model describing the toxicokinetics of BaP and 3-OHBaP specifically.

The modeling approach consisted of developing a biologically-based kinetic model in rats that captures essential biological determinants of the kinetics of BaP and its key biomarker of exposure, 3-OHBaP, for later extension to humans. It emphasizes on mass-balance and

77

uses available *in vivo* time courses of BaP and 3-OHBaP in blood, key tissues and excreta for the determination of model parameters, without the need for detailed determination of anatomical or *in vitro*-based physiological parameters. In the determination of model parameter values, extensive use is made of the different time scales over which the various biological processes occur. Regrouping tissues according to these specific time scales simplified the determination of the key parameters governing the overall model kinetics.

4.3 Materials and methods

4.3.1 General modeling approach

A toxicokinetic model has been developed where BaP and 3-OHBaP in blood, tissues and excretas were represented as compartments. The conceptual representation of the model was based on the analysis of the kinetic behavior experimentally observed from data collected in our laboratory (Bouchard and Viau, 1996; Marie et al., 2010). The time evolution of BaP and 3-OHBaP amounts in each compartment of the model was then mathematically described by a system of coupled differential equations.

4.3.2 Experimental data used for model development

The data of Marie et al. (2010) on the time profile of BaP and 3-OHBaP in blood, tissues (liver, kidney, lung, adipose tissues and skin), and excreta (urine, feces) of male Sprague-Dawley rats over a 72-h period following an intravenous injection of 40 μmol/kg bw of BaP served as a basis for the development of the toxicokinetic model for BaP and 3-OHBaP (collection times = 2, 4, 8, 16, 24, 33, 48, 72 h post-injection).

4.3.3 Model representation

Figure 4-2 illustrates the conceptual model of the kinetics of BaP and its 3-OHBaP biomarker in blood, tissues, and excreta. Symbols and abbreviations are described in Table 4-1. In the model, blood and key organs contributing significantly to the absorption, distribution, and retention of BaP or 3-OHBaP, or otherwise to the metabolism of BaP into 3-OHBaP or other metabolites, were represented as compartments as well as urinary and faecal elimination routes. The rates of change in the amounts of BaP or 3-OHBaP ($dX_i(t)/dt$) in a given compartment (on a molar basis) were then expressed by the difference between incoming and outgoing rates from the compartment. Transfer rates between compartments represented either the physical transfer of BaP or 3-OHBaP from one organ to the other, or the absorption rate at a site-of-entry compartment or otherwise the biotransformation rate of BaP into 3-OHBaP or into other metabolites in a metabolizing organ (on a molar basis). Resolution of the differential equations simulating the kinetics of BaP and 3-OHBaP in the body generated the mathematical functions ($X_i(t)$) describing the time profile of these molecules in the different model compartments.

The model was built while ensuring conservation of mass (in moles), hence at all times, the dose was always equal to the sum of burdens in the different compartments (parent compounds and metabolites) as well as those accumulated in excreta since exposure.

4.3.4 Determination of model parameter values

The parameter values estimated in the kinetic model were the inter-compartment transfer rates of BaP and 3-OHBaP, the metabolism rates of BaP into 3-OHBaP and other metabolites, the elimination rates of BaP and 3-OHBaP, and their fractional partition in tissues in equilibrium with blood (see Table 4-1 for the different model parameters).

Parameter values were established by best-fit adjustments of the analytical solution of the differential equations to the experimental kinetic data available of Marie et al. (2010) on the time profiles of BaP and 3-OHBaP in blood, tissues, and excreta following intravenous injection of BaP). The software Mathematica 7.0 from Wolfram Research, Inc. (Champaign, Illinois) was used to analytically solve the system of differential equations. It is important to note that several procedures exist to best-fit general analytical functions to data sets. For fitting, the algorithm FindFit (included in Mathematica) was used, which essentially reproduces a least square minimization.

To simplify differential equations and allow a first estimate of parameter values, extensive use was made of the different time scales during which the different biological processes occur (e.g., time of absorption, distribution, metabolism, and excretion). These different time scales render the use of the quasi steady state approximation (QSSA) possible (Segel 1988; Segel and Slemrod 1989). Briefly, QSSA predicts that a compartment X reaches a kinetic equilibrium with its feeding compartment when the output rate is significantly more rapid than the time variations in its feeding source. In these conditions, the burden in the compartment X is always proportional to that of the source compartment. Therefore, if the burden of the feeding compartment varies as a function of time, that of the compartment X will vary such that the ratio of burdens in compartment X to that of the feeding compartment remains constant. This phenomenon is very frequent in kinetics of drugs and toxic substances.

With these considerations, over specific time periods, the experimental data can be expressed by a single exponential function. For example, in the case of the BaP blood compartment, the first couple of hours are primarily governed by the rates of absorption; between 6 and 24 h post-dosing, blood time course of BaP is determined by distribution of BaP from blood to tissues; at times >24 h, all the BaP has been distributed to tissues, hence, the blood time course of BaP is simply driven by the slow feeding from storage tissues, hence lungs and adipose tissues.

As another example, the analytical solution describing the time course of BaP in the lung (LU(t)) is proportional to (see Table 4-1 for the description of symbols and abbreviations):

$$LU(t) \propto \frac{2DK_{BLU}}{\sqrt{(K_{LGI} + K_{KU} + K_{BAT} + K_{BLU} + K_{LUB})^2 - 4K_{LUB}(K_{LGI} + K_{KU} + K_{BAT})}}$$
$$\left(e^{-\frac{t}{2}\left[K_{LGI}+K_{KU}+K_{BAT}+K_{BLU}+K_{LUB} + \sqrt{(K_{LGI}+K_{KU}+K_{BAT}+K_{BLU}+K_{LUB})^2 - 4K_{LUB}(K_{LGI}+K_{KU}+K_{BAT})} \right]} \right.$$
$$\left. - e^{-\frac{t}{2}\left[K_{LGI}+K_{KU}+K_{BAT}+K_{BLU}+K_{LUB} - \sqrt{(K_{LGI}+K_{KU}+K_{BAT}+K_{BLU}+K_{LUB})^2 - 4K_{LUB}(K_{LGI}+K_{KU}+K_{BAT})} \right]} \right)$$

(4-1)

However, at times >24 h, this function reduces to the single exponential:

$$LU(t) \approx \frac{2DK_{BLU}}{\sqrt{(K_{LGI}+K_{KU}+K_{BAT}+K_{BLU}+K_{LUB})^2}} \left(e^{-tK_{LUB}} - 1 \right)$$ (4-2)

With those initial values fixed, an iteration procedure allowing a one-by-one re-adjustment of each parameter values for several cycles (at least a hundred for each parameter). As a first step, parameters describing the kinetics of BaP were fixed and, as a second step, an optimization of parameter values related to 3-OHBaP kinetics was performed. The algorithm started by fitting the critical transfer rates governing the kinetics while the other values were kept constant. Once an iteration cycle was completed, another fitting cycle began with the newly computed set of parameters. This recursive process was carried out for several cycles until the resulting variations in the fitted values were negligible. Each of these best-fittings was based on the least-square minimization method included in Mathematica and on

81

common linear regression (for an extensive explanation on the algorithm, please see Weisstein 2011).

4.3.5 Experimental data used for model validation

The data of Bouchard and Viau (1996) on the time course of 3-OHBaP in the urine of male Sprague-Dawley rats over a 90-h period following an intravenous injection of 40 μmol/kg bw of BaP were used for model validation (collection times = 2, 4, 6, 8, 10, 12, 18, 24, 30, 42, 48, 54, 66, 72, 78, 90 h post-injection of BaP). In order to define the goodness of fit of our model, the Pearson's chi-square test was used.

The data of Lee et al. (1996) on the time course of 3-OHBaP in the urine of adult male Sprague-Dawley rats over a 96-h period following an intraperitoneal injection of 20 mg/kg bw of BaP were also used for model validation (at 4-12 h intervals post-injection of BaP).

As a preliminary assessment of the use of the modeling approach for human biomonitoring, the available published human data of Lafontaine et al. (2004) on the time course of 3-OHBaP in the urine of a worker over a 48-h period following an occupational respiratory exposure to BaP were further used. These authors collected all urine voided starting with pre-shift beginning of workweek urine samples following at least a 36-h period without occupational exposure and ending with beginning-of-shift urine sample of the third working day. Parameter values were kept as determined in animals except those governing the overall excretion kinetics of the key biomarker of exposure 3-OHBaP.

4.3.6 Model simulations

The analytical resolution of the systems of equations, described in the previous section, was a required step to fix the transfer values of the toxicokinetic model, by fitting experimental data of Marie et al. (2010). Once the parameter values of the model were determined in Mathematica, the system of differential equations representing the complete model of the kinetics of BaP and 3-OHBaP was numerically solved in MathCad 14 (PTC (Parametric Technology Corporation), Needham, MA, USA) using the Adams-Bashforth-Moulton method in order to reproduce the experimental data of Marie et al. (2010), Bouchard and Viau (1996), Lee et al. (2003), and Lafontaine et al. (2004).

4.4 Results

Table 4-2 presents parameter values of the model and Table 4-3 blood and tissue half-lives calculated from model parameters; this single set of parameters was used in all data simulations. Figure 4-3, Figure 4-4 and Figure 4-5 show that the model reproduced closely the data of Marie et al. (2010) on the time courses of BaP and 3-OHBaP in blood, tissues, urine, and feces of rats exposed intravenously to BaP.

The excellent correspondence of our mathematical model with experimental data has corroborated our principal hypotheses about the BaP and 3-OHBaP kinetics, namely: a) BaP and 3-OHBaP are rapidly distributed (see Figure 4-3, Figure 4-4 and Figure 4-5), b) most organs monitored are in kinetic equilibrium with blood (see Figure 4-3b,c), c) liver and lungs are two biotransformation organs for BaP, d) lipidic components in organs act as storage compartments for BaP and to a certain extent for 3-OHBaP, e) a build-up of 3-OHBaP is found in the kidneys (see Fig. Figure 4-4c), and f) 3-OHBaP is only a quantitatively minor metabolite of BaP (see Figure 4-5).

Overall elimination half-life of BaP and 3-OHBaP from blood (and tissues in equilibrium with blood) was found to be 3.65 and 2.26 h respectively, which represents a rapid tissue distribution, biotransformation, and elimination of both molecules (see Table 4-2). Five different processes contribute to this total elimination rate of BaP and 3-OHBaP: biotransformation in the liver, elimination through the gastrointestinal tract, renal clearance, distribution to lipidic components, and finally distribution to the lungs (see Figure 4-2). Of all the preceding mechanisms, the most important contribution to the overall elimination rate of BaP from blood is liver biotransformation (72.6%). On the other hand, elimination of 3-OHBaP through the gastro-intestinal tract contributes mainly to 3-OHBaP overall clearance from blood (62.9%).

Since distribution of BaP and 3-OHBaP in the body is very rapid, some tissues are found to reach a kinetic equilibrium with blood. In the case of BaP, this assumption was also very well corroborated by our model in which blood, skin, liver, kidneys, and other non-observed tissue were described by a fraction of the analytical function $\overline{B}(t)$, which represents the sum of BaP levels in blood and in tissues where a kinetic equilibrium with blood is rapidly reached (see Table 4-2). Consequently, to obtain the function describing BaP present in blood, liver, skin, and kidneys, this function simply has to be multiplied by the fractional partition for the compartment of interest: $B(t)=f_B\,\overline{B}(t)$, $L(t)=f_L\,\overline{B}(t)$, $S(t)=f_S\,\overline{B}(t)$ and $K(t)=f_K\,\overline{B}(t)$. The blood compartment only accounts for 0.914% of the total of BaP found in blood and all tissues in equilibrium with blood. In a similar manner, kidneys just receive 0.935% of total BaP in blood and tissues in equilibrium with blood. Among the studied tissues in equilibrium with blood, those showing highest partitioning of BaP were the skin (6.27% of total BaP in blood and tissues in equilibrium with blood) and the liver (12.8%). On the other hand, for 3-OHBaP, only the skin and the liver were found to be in equilibrium with the blood compartment with corresponding partitioning of 2.67% for blood, 4.12% for skin, and 5.07% for liver. The distribution is more uniform for 3-OHBaP given that it is less lipophilic than its parent compound.

Conversely, lungs and adipose tissues were not observed to be in equilibrium with blood (at initial times) and were found to present the highest fractions of BaP dose. Firstly, a relatively rapid tissue uptake is evident from their measured time profiles ($t_{1/2} = 9.72$ min for lungs and $t_{1/2} = 43.7$ min for adipose tissues). As shown in Figure 4-2, uptake of BaP in the lungs or adipose tissues is represented by the transfer rate from the blood compartment to the lungs or adipose tissues, K_{BLU} and K_{BAT}, respectively. These parameters, at first order, simply have the control over the rate of uptake and total amounts of BaP found in lungs or adipose tissues.

Subsequently, the parallel time course of BaP in the lungs and adipose tissues indicate a similar retention mechanism once BaP reaches both these tissues ($t_{1/2} = 29.3$ h for lungs and

85

$t_{1/2} = 27.2$ h for adipose tissues). The slopes of the time course curves of 3-OHBaP in these tissues are also similar to those of BaP, which can be explained by the same biological determinants governing the kinetics. However, 3-OHBaP is much less retained in lungs because the amounts of 3-OHBaP observed were significantly lower than to those of adipose tissues. Two hours post-injection, 2.47% of BaP were found in adipose tissues and 17.0% of BaP in lungs while only 0.0230% and 0.0241% of 3-OHBaP were observed in both lungs and adipose tissues, respectively. The elimination rate of BaP from the lung (K_{LU} = $K_{LUB}+K_{LUlu}+K_{LUOLU} = 2.37\times10^{-2}$ h^{-1}) and the transfer rates of BaP from adipose tissues to blood ($K_{ATB} = 2.55\times10^{-2}$ h^{-1}) are the most sensitive in the whole kinetics, for if they are too large, storage will no longer occur. Conversely, if they are too small, the first phase of rapid distribution of BaP from blood to lungs or adipose tissues will simply consist of an endless accumulation of BaP and the kinetics would no longer be driven by the parent compound.

Concerning the kidney, in the BaP case, it responds as any other organ in kinetic equilibrium with blood. However, this is not the case for 3-OHBaP for which a small built-up in the kidney is observed. Our toxicokinetic model established a smaller elimination rate of 3-OHBaP from the kidneys (either by excretion to urine, k_{ku}, or by reabsorption into blood, k_{kb} ($t_{1/2} = 4.52$ h) compared to the rate of transfer from blood to kidney, k_{bk} ($t_{1/2} = 28.8$ min). This difference in the rates is responsible for the maximum 3-OHBaP kidney values observed at around 8 h. The values of the parameters k_{bk}, k_{kb} and k_{ku}, determines maximum levels of 3-OHBaP in the kidneys and time to peak-levels. The later value, k_{ku} is also fixed by the fraction of injected BaP recovered overall in urine as 3-OHBaP.

Regarding biotransformation, our model accounts for liver and lung metabolism. In the liver and lung, total elimination of BaP occurs mainly through metabolism (99.5% of BaP reaching the liver and 86.5% of lung BaP being metabolized). According to our model, 3-OHBaP itself is also partly metabolized (26.5% of 3-OHBaP reaching the liver being metabolized). Liver metabolism of BaP to metabolites (3-OHBaP and other metabolites) as well as biotransformation of 3-OHBaP itself had to be considered relatively fast to provide a close approximation to the set of data, but limited by the rate of effective distribution of BaP from blood and tissues in equilibrium with blood in the case of the liver metabolism. Lung metabolism of BaP into 3-OHBaP and other metabolites was found to be slow ($K_{LUlu}+K_{LUOLU} = 2.05 \times 10^{-2}$ h^{-1}) compared to that of the liver, but it was faster than the transfer of BaP from lung to the blood stream ($K_{LUB} = 3.19 \times 10^{-3}$ h^{-1}). There are thus two different parameters representing biotransformation process in both liver and lung. In the lung compartment, BaP is converted into 3-OHBaP at a low K_{Lulu} rate of 3.91×10^{-3} h^{-1} ($t_{\frac{1}{2}} =$ 7.39 d) and into other metabolites at the K_{LUOLU} rate of 1.66×10^{-2} h^{-1} ($t_{\frac{1}{2}} = 41.8$ h). In the liver, the corresponding rates are much more rapid than those of the lungs with K_{LI} rate of 2.09×10^{-1} h^{-1} ($t_{\frac{1}{2}} = 3.32$ h) and K_{LOL} rate of 8.68×10^{-1} h^{-1} ($t_{\frac{1}{2}} = 47.9$ min).

It was then verified that, with the parameters described previously, the model was able to closely reproduce the cumulative excretions in urine and feces for both substances (see coefficients of goodness of fit in Table 4-3), as expected given the direct link with metabolites levels in biological matrices measured for a given internal BaP dose. Simulations were in excellent accordance with the observed cumulative excretion time courses, especially considering the fact that measurements are not "real" cumulative excretions from a single rat but rather are collections from different rats sacrificed at different time points.

In the particular case of the gastrointestinal tract, it has been modeled as separate sub-compartments to describe the kinetics of both BaP and 3-OHBaP. The first segment (GI$_1$ for BaP and gi$_1$ for 3-OHBaP) is the one that effectively represents the kinetics of BaP and 3-OHBaP in the gastrointestinal tract and therefore receiving liver burdens and representing possible enterohepatic recycling of 3-OHBaP. By definition, the first segments describe different kinetics for BaP and 3-OHBaP. Transfer rate values from liver to the GI$_1$ segment and from the GI$_1$ to GI$_2$ segments are $K_{LGI} = 5.89 \times 10^{-3}$ h^{-1} and $K_{GIGI} = 4.96 \times 10^{-2}$ h^{-1} for the

BaP compartment and $k_{lgi} = 3.81$ h^{-1} and $k_{gigi} = 7.39 \times 10^{-2}$ h^{-1} for 3-OHBaP compartment while enterohepatic recycling of 3-OHBaP from the GI back to the liver is represented by the k_{gil} rate of 1.77×10^{-2} h^{-1}. The second segment (GI_2 for BaP and gi_2 for 3-OHBaP) simply represents a compartment which delays faecal excretion with a half-life of 5.55 h. This segment acts as a transition delay to account for the gastrointestinal tract transit time. For this reason, elimination rate values associated with this second segment (K_{GIF} and k_{gif}) were set to be equal for BaP and 3-OHBaP.

With the parameter values determined from the rat data of Marie et al. (2010), the model was found to provide an excellent approximation ($\chi^2 = 0.323$) of the experimental data of Bouchard and Viau (1996) on the excretion time course of 3-OHBaP in the urine of rats intravenously exposed to BaP (Figure 4-6), with no a posteriori adjustment of parameters. The p-value corresponding to five degrees of freedom in the experimental data (the data set reported in Bouchard and Viau (1996) was obtained from a sample of five rats) was calculated to be very small, 0.00281.

The model also reproduced closely the experimental data of Lee et al. (2003) (Figure 4-7). However, this is based on a visual assessment given that these authors did not report the standard deviation of their experimental data such that the corresponding chi-square value could not be computed. Furthermore, since the experimental data of Lee et al. (2003) was obtained in rats exposed intraperitoneally (ip), simulation was performed with a delay of 10.5 h to account for differences between the iv and ip routes of exposure.

On the basis of the rat model and animal-to-human adjustment of parameter values governing the overall excretion kinetics of the key biomarker of exposure 3-OHBaP, the time course of 3-OHBaP in the urine of a worker over a 48-h period following an occupational respiratory exposure to BaP was simulated. A good approximation of the available published human time course data was obtained by keeping unchanged all the

88

parameters established from rata data except for the transfer rates of both BaP and 3-OHBaP from adipose tissues or lung to blood (Figure 4-8). The latter transfer rates were shown to determine the excretion kinetics. Animal-to-human extrapolation of the transfer rates of both BaP and 3-OHBaP from adipose tissues or lung to blood was conducted using the following factor:

$$K_{ATB}^{human} = \frac{V_{fat}^{rat} P_{fat}^{rat} Q_{fat}^{human}}{V_{fat}^{human} P_{fat}^{human} Q_{fat}^{rat}} K_{ATB}^{rat} \tag{4-3}$$

$$K_{LUB}^{human} = \frac{V_{lungs}^{rat} P_{lungs}^{rat} Q_{lungs}^{human}}{V_{lungs}^{human} P_{lungs}^{human} Q_{lungs}^{rat}} K_{LUB}^{rat} \tag{4-4}$$

where K_{ATB}^{human} and K_{ATB}^{rat} are the transfer rates of BaP from adipose tissues to blood in humans and rats, respectively, Q_{fat}^{human} and Q_{fat}^{rat} represent the regional blood flow rates in humans and rats, respectively, V_{fat}^{human} and V_{fat}^{rat} are the volumes of adipose tissues in humans and rats, respectively, and P_{fat} is the adipose tissues-blood partition coefficient, which is taken to be equal in humans and rats. K_{LUB}^{human} and K_{LUB}^{rat} are the transfer rates of BaP from lung to blood in humans and rats, respectively, Q_{lung}^{human} and Q_{lung}^{rat} represent the regional blood flow rate in humans and rats, respectively, V_{lung}^{human} and V_{lung}^{rat} are the volumes of lung tissues in humans and rats, respectively, and P_{lung} is the lung tissue-blood partition coefficient, which is also taken to be equal in humans and rats. The same type of extrapolation was done for the corresponding 3-OHBaP parameter values. Simulation conducted with these adjusted human model parameters was very accurate without the need for adjustment of any other parameter.

89

4.5 Discussion

This study allowed a better understanding of the kinetics of 3-OHBaP, a potential key biomarker of exposure to BaP. It enabled to relate an internal dose of BaP to its time course in key tissues and accessible biological fluids (such a blood and urine) along with that of its biomarker 3-OHBaP. The model provided a very good approximation of the experimental time course data of Marie et al. (2010), Bouchard and Viau (1996), Lee et al. (2003), and Lafontaine et al. (2004).

4.5.1 Kinetics of BaP and model simulations

According to model simulations, the main biological processes governing the kinetics of BaP are the uptake of BaP in the lipidic components of the lung (such as membrane lipids) and in adipose tissues combined to a slow metabolism of BaP in the lung compared to the liver as confirmed by experimental data (Mitchell, 1983; Prough et al., 1979). The lipophilic properties of BaP and slow lung metabolism thus appear to explain in large part its uptake in lipidic components and parallel slow elimination time course of BaP in lung and adipose tissues compared to other tissues, such as the liver. In the lung, this is combined to the fact that, following intravenous injection, it is the first organ, after the heart, to receive total intravenous dose. After an oral administration of BaP in rats, Ramesh et al. (2001) found no such accumulation in lung; in the latter study, BaP levels were similar in lung and liver while plasma was more concentrated in BaP.

In the liver, skin and kidney compartments of the model, levels of BaP are in kinetic equilibrium with those of BaP in the blood compartment, such that they evolve in parallel, as experimentally observed. The more rapid elimination rate of BaP in blood and these tissues reflects renal clearance and biliary elimination rate of rapidly distributed and

90

metabolized BaP, while the slower elimination phase is driven by the rate-limiting elimination of BaP stored in lung components and adipose tissues. This second stage of slower elimination drives the whole kinetics as soon as initial tissue distribution and uptake are completed.

The current toxicokinetic model for BaP suggests that total amounts of BaP in the lung as a function of time is determined not only by the lipid affinity, but also by some other storage process, otherwise adipose tissues would have presented larger amounts of BaP. The affinity difference between these two compartments could be regarded as a rough indicator of protein binding contribution (about 77.8%) in the lungs, supposing that lipid components of the lungs are comparable to those of adipose tissues measured, as follows:

$$Protein\ Binding \sim \frac{K_{BLU} - K_{BAT}}{K_{BLU}}\ . \tag{4-5}$$

In the liver, BaP affinity for proteins is accounted for by its large partition coefficient ($f_L = 12.8\%$), compared to those of the other organs, which are in kinetic equilibrium with blood. This suggests again that there is an affinity mechanism that leads the higher BaP levels in the liver compared to other organs. These results are all in accordance with published in vitro studies showing significant binding of BaP to Ah receptors along with P450 1A1 induction in liver and lung (Shimada et al., 2002).

With regard to the metabolism of BaP, as mentioned previously, model also suggests that liver metabolism of BaP was significantly higher than lung metabolism ($t_{½LI} = 3.32$ h and $t_{½Lulu} = 7.34$ d, respectively). This is in complete agreement with experimentally observed slow disappearance of BaP from the lungs compared to the rapid disappearance from the liver. Furthermore, the model assessed that lung metabolism to 3-OHBaP is much less important than biotransformation to other BaP metabolites.

Also according to model simulations, no saturation in lung and liver metabolism of BaP was apparent at the 40 µmol/kg intravenous dose administered to rats. BaP time profile in the lung, where highest BaP concentrations were observed, was very similar to that of adipose tissues, whose levels were an order of magnitude smaller. This kind of kinetics is an indication that no saturation mechanism is present at the injected dose.

4.5.2 Kinetics of 3-OHBaP and model simulations

With regard to 3-OHBaP kinetics, the model suggests that the terminal elimination phase in blood, liver, skin and lung is driven in large part by the elimination rate of BaP stored in lung, adipose tissue or other body lipid components. There is also definitely a certain contribution of 3-OHBaP itself stored in adipose tissues and tissue lipid components. As observed for BaP, there must be an affinity of the metabolite 3-OHBaP for lipids. The accumulation of 3-OHBaP in adipose tissues hence reflects the lipophilicity of the metabolite itself even after monohydroxylation of the parent compound.

In the toxicokinetic model, 3-OHBaP and BaP in the gastrointestinal tract have further been represented by two sub-compartments. As mentioned in the Results section, the first segment represents the actual kinetics of the gastrointestinal tract while the second simply delays appearance in faeces with a half-life of 5.55 h for both molecules. Differences are apparent in the elimination kinetics of BaP and 3-OHBaP. First, the liver transfers only very small amounts of BaP to the gastrointestinal tract with a half-life of 4.91 d while large amounts of 3-OHBaP are secreted with a $t_{1/2}$ of 10.9 min. In addition to this difference in the GI tract uptake of the two compounds (three orders of magnitude), elimination from the GI also differs. In the case of BaP, faecal excretion is limited with a $t_{1/2}$ of 14.0 h. In the case of 3-OHBaP, faecal excretion is substantial; overall elimination of 3-OHBaP from the GI was found to occur with a $t_{1/2}$ of 7.57 h.

92

The model also accounted for enterohepatic recycling of 3-OHBaP by including a transfer of 3-OHBaP (kgil) from the first segment of the gastrointestinal tract back to the liver compartment. Significant enterohepatic recycling of BaP metabolites has also been described in a study in cannulated rats (Chipman et al., 1981). The same type of recycling can also occur for BaP (Chipman et al., 1981); nevertheless, contribution of this process to the overall excretion kinetics is negligible, given the very small fraction of internal BaP dose observed as BaP in faeces (0.397%) compared to that of 3-OHBaP (12.9%) (i.e., two orders of magnitude apart). Our mathematical model showed that this recycling mechanism of 3-OHBaP has a noticeable impact on the kinetics of distribution and elimination of this metabolite and therefore contributes significantly to the overall cumulative amounts of 3-OHBaP observed in urine (19.3% of the available 3-OHBaP is reabsorbed according to the model).

As for the atypical time course of 3-OHBaP in urine experimentally documented by Bouchard and Viau (1996) and Lee et al. (2003), it is determined by its kidney uptake and slow output rate according to the model. Uptake of 3-OHBaP in most tissues (i.e., liver, blood, skin, lungs and adipose tissues) is governed by the fast metabolism rate of BaP in the liver ($t_{1/2}$ = 3.32 h). However, in kidney, maximum levels of 3-OHBaP are reached only around 8 h after injection as compared to 4 h post-dosing in the liver, although large amounts of metabolites are found in kidney compared to other organs in equilibrium with blood. Such kidney profile can exist only if there is a relatively rapid uptake in kidneys ($t_{1/2}$ = 27.8 min) but a slower elimination ($t_{1/2}$ = 4.52 h), either by excretion of 3-OHBaP in urine or its transfer back to blood. This built-up phenomenon of this metabolite in this organ may be explained by a delayed active tubular secretion of the conjugated form of this metabolite in proximal tubules (Gosselin et al., 2005).

93

4.5.3 Comparison with other published models

Roth and Vinegar (1990) developed a PBPK model to describe the kinetics of the parent compound BaP following intraarterial injection. Metabolism was considered to occur in both the lung and liver and binding was incorporated in the liver and lung compartments. Parametric values are however not reported in this study.

Furthermore, Bevan and Weyand (1988) developed a compartmental model of the distribution of total radioactivity in male Sprague-Dawley rats following intratracheal instillation of 3H-BaP. Compartments included blood, liver, lung, intestines (including its contents) and carcasses (the latter of which were modeled as the sum of two compartments) along with undefined additional compartments introduced in the model to account for a delay in urinary excretion of radioactivity or in the transfer of radioactivity from liver to the intestinal compartments through biliary secretion. The delay in urinary excretion of blood radioactivity introduced in the model of Bevan and Weyand (1988) is compatible with the 3-OHBaP kidney built-up considered in our model. As in our model, these authors also accounted for a significant enterohepatic recycling of BaP metabolites to provide an adequate fit to experimental data. Although these models provided valuable information on the kinetics of BaP and total radioactivity, they did not focus on the kinetic modeling of 3-OHBaP as a potential key biomarker of exposure to BaP.

Overall, this study succeeded in developing a toxicokinetic model for BaP and 3-OHBaP, which provided a close match to a large set of experimental time course data in several biological matrices of exposed rats. This modeling provided new insights into the mechanistic determinants of 3-OHBaP kinetics that can serve for a better understanding and use of 3-OHBaP biomonitoring data.

4.6 Appendix: Linear differential equations for each model compartment

The mathematical representation of 4-2 is given by the following system of differential equations (see 4-1 for definitions of symbols and abbreviations):

4.6.1 Kinetics of BaP

$$\frac{\partial \bar{B}(t)}{\partial t} + [K_{KU}f_K + (K_{LGI} + K_{LOL} + K_{LI})f_L + (K_{BAT} + K_{BLU})f_B]\bar{B}(t) = D_0\delta(t) +$$

$$K_{ATB}AT(t) + K_{LUB}LU(t) \tag{4-6}$$

$$\frac{\partial AT(t)}{\partial t} + K_{ATB}AT(t) = K_{BAT}f_B\bar{B}(t) \tag{4-7}$$

$$\frac{\partial LU(t)}{\partial t} + K_{LUB}LU(t) = K_{BLU}f_B\bar{B}(t) \tag{4-8}$$

$$\frac{\partial GI_1(t)}{\partial t} + K_{GIGI}GI_1(t) = K_{LGI}f_L\bar{B}(t) \tag{4-9}$$

$$\frac{\partial GI_2(t)}{\partial t} + K_{GIF}GI_2(t) = K_{GIGI}GI_1(t) \tag{4-10}$$

$$\frac{\partial \Gamma(t)}{\partial t} = K_{GIF}GI_2(t) \tag{4-11}$$

$$\frac{\partial U(t)}{\partial t} = K_{KU}f_K\bar{B}(t) \tag{4-12}$$

$$B(t) = f_B\bar{B}(t) \tag{4-13}$$

$$L(t) = f_L\bar{B}(t) \tag{4-14}$$

$$S(t) = f_S\bar{B}(t) \tag{4-15}$$

$$K(t) = f_K\bar{B}(t) \tag{4-16}$$

4.6.2 Kinetics of 3-OHBaP metabolites

$$\frac{\partial \bar{b}(t)}{\partial t} + \left[\left(k_{lgi} + k_{lol} \right) f_l + \left(k_{bat} + k_{blu} + k_{bk} \right) f_b \right] \bar{b}(t) = K_{Ll} f_L \bar{B}(t) + k_{atb} at(t) +$$

$$k_{lub} lu(t) + k_{kb} k(t) + k_{gib} gi_1(t)$$

$$(4\text{-}17)$$

$$\frac{\partial at(t)}{\partial t} + k_{atb} at(t) = k_{bat} f_b \bar{b}(t) \tag{4-18}$$

$$\frac{\partial lu(t)}{\partial t} + k_{lub} lu(t) = k_{blu} f_b \bar{b}(t) + K_{LUlu} LU(t) \tag{4-19}$$

$$\frac{\partial gi_1(t)}{\partial t} + \left(k_{gif} + k_{gib} \right) gi_1(t) = k_{lgi} f_l \bar{b}(t) \tag{4-20}$$

$$\frac{\partial gi_2(t)}{\partial t} + k_{gigi} gi_2(t) = k_{gif} gi_1(t) \tag{4-21}$$

$$\frac{\partial f(t)}{\partial t} = k_{gif} gi_2(t) \tag{4-22}$$

$$\frac{\partial k(t)}{\partial t} + (k_{ku} + k_{kb}) k(t) = k_{bk} f_b \bar{b}(t) \tag{4-23}$$

$$\frac{\partial u(t)}{\partial t} = k_{ku} k(t) \tag{4-24}$$

$$b(t) = f_b \bar{b}(t) \tag{4-25}$$

$$l(t) = f_l \bar{b}(t) \tag{4-26}$$

$$s(t) = f_s \bar{b}(t) \tag{4-27}$$

4.7 Funding

This work was supported by the Agence nationale de sécurité sanitaire de l'alimentation, de l'environnement et du travail (ANSES) [Grant number 2009-CRD-24].

4.8 References

Agency for Toxic Substances and Disease Registry (ATSDR). 1995. Toxicological profile for Polycyclic Aromatic Hydrocarbons (PAHs). Atlanta, GA: U.S. Department of Health and Human Services, Public Health Service.

Angerer, J., Ewers, U., and Wilhelm, M. (2007). Human biomonitoring: state of the art. Int. J. Hyg. Environ. Health 210, 201-228.

Bevan, D.R., and Weyand, E.H. (1988). Compartmental analysis of the disposition of benzo(a)pyrene in rats. Carcinogenesis 9, 2027-2032.

Bouchard, M., and Viau, C. (1996). Urinary excretion kinetics of pyrene and benzo(a)pyrene metabolites following intravenous administration of the parent compounds or the metabolites. Toxicol. Appl. Pharmacol. 139, 301-309.

Bouchard, M., and Viau, C. (1999). Urinary 1-hydroxypyrene as a biomarker of exposure to polycyclic aromatic hydrocarbons: biological monitoring strategies and methodology for determining biological exposure indices for various work environments. Biomarkers 4, 159 - 187.

Bouchard, M., Normandin, L., Gagnon, F., Viau, C., Dumas, P., Gaudreau, E., and Tremblay, C. (2009). Repeated measures of validated and novel biomarkers of exposure to polycyclic aromatic hydrocarbons in individuals living near an aluminum plant in Quebec, Canada. J. Toxicol. Environ. Health 72, 1280-1295.

Chipman, J.K., Hirom, P.C., Frost, G.S., and Millburn, P. (1981). The biliary excretion and enterohepatic circulation of benzo(a)pyrene and its metabolites in the rat. Biochem. Pharmacol. 30, 937-944.

Forster, K., Preuss, R., Rossbach, B., Bruning, T., Angerer, J., and Simon, P. (2008). 3-Hydroxybenzo[a]pyrene in the urine of workers with occupational exposure to polycyclic aromatic hydrocarbons in different industries. Occup. Environ. Med. 65, 224-229.

Gendre, C., Lafontaine, M., Morele, Y., Payan, J.P., and Simon, P. (2002). Relationship between urinary levels of 1-hydroxypyrene and 3-hydroxybenz[a]pyrene for workers exposed to polycyclic aromatic hydrocarbons. Polycycl. Aromat. Compd. 22, 761-769.

Gendre, C., Lafontaine, M., Delsaut, P., and Simon, P. (2004). Exposure to polycyclic aromatic hydrocarbons and excretion of urinary 3-hydroxybenzo[a]pyrene (3-OHBaP): assessment of an appropriate sampling time. Polycycl. Aromat. Compd. 24, 433-439.

Gosselin, N.H., Bouchard, M., Brunet, R.C., Dumoulin, M.J., and Carrier, G. (2005). Toxicokinetic modeling of parathion and its metabolites in humans for the determination of biological reference values. Toxicol. Mech. Meth. 15, 33-52.

Gupta, P., D.K. Banerjee, S.K. Bhargava, R. Kaul and V. R. Shankar (1993). Prevalence of impaired lung function in rubber manufacturing factory workers exposed to benzo(a)pyrene and respirable particulate matter. Indoor Environ., 2:26-31.

Jacob, J., and Seidel, A. (2002). Biomonitoring of polycyclic aromatic hydrocarbons in human urine. J. Chromatogr. B Analyt. Technol. Biomed. Life Sci. 778, 31-47.

Jongeneelen, F.J., Anzion, R.B.M., and Henderson, P.T. (1987). Determination of hydroxylated metabolites of polycyclic aromatic hydrocarbons in urine. J. Chromatogr. 413, 227 232.

Lafontaine, M., Gendre, C., Delsaut, P., and Simon, P. (2004). Urinary 3-htydroxybenzo[a]pyrene as a biomarker of exposure to polycyclic aromatic hydrocarbons: an approach for determining a biological limite value. Polycycl. Aromat. Compd. 24, 441-450.

Lee, W., Shin, H.S., Hong, J.E., Pyo, H., and Kim, Y. (2003). Studies on the analysis of benzo(a)pyrene and its metabolites in biological samples by using high performance liquid chromatography/fluorescence detection and gas chromatography/mass spectrometry. Bull. Korean Chem. Soc. 24, 559-565.

Maître, A., Badouard, C., Pauthier, E., Marques, M., Le Gall, L., and Stoklov, M. (2008). Cartographie d'exposition professionnelle aux hydrocarbures aromatiques polycycliques (HAP) : intérêt de la fiche de renseignements sur l'activité professionnelle. Arch. Mal. Prof. 69, 328-329.

Marie, C., Bouchard, M., Heredia-Ortiz, R., Viau, C., and Maître, A. (2010). A toxicokinetic study to elucidate 3-hydroxybenzo(a)pyrene atypical urinary excretion profile following intravenous injection of benzo(a)pyrene in rats. J. Appl. Toxicol. 30, 402-410.

Mitchell, C.E. (1982). Distribution and retention of benzo(A)pyrene in rats after inhalation. Toxicol. Lett. 11, 35-42.

Mitchell, CE. (1983). The metabolic fate of benzo[a]pyrene in rats after inhalation. Toxicology 28, 65-73.

Moir, D., Viau, A., Chu, I., Withey, J., and McMullen, E. (1998). Pharmacokinetics of benzo[a]pyrene in the rat. J. Toxicol. Environ. Health A 53, 507-530.

Prough, R.A., Patrizi, V.W., Okita, R.T., Masters, B.S., Jakobsson, S.W. (1979). Characteristics of benzo(a)pyrene metabolism by kidney, liver, and lung microsomal fractions from rodents and humans. Cancer Res. 39, 1199-1206.

Ramesh, A., Inyang, F., Hood, D.B., Archibong, A.E., Knuckles, M.E., and Nyanda, A.M. (2001). Metabolism, bioavailability, and toxicokinetics of benzo(α)pyrene in F-344 rats following oral administration. Exp. Toxicol. Pathol. 53, 275-290.

Roth, R.A., and Vinegar, A. (1990). Action by the lungs on circulating xenobiotic agents, with a case study of physiologically based pharmacokinetic modeling of benzo(a)pyrene disposition. Pharmacol. Ther. 48, 143-155. Segel, L.A. (1988). On the validity of the Steady State Approximation of enzyme kinetics. Bull. Math. Biol. 50, 579-593.

Segel, L.A., and Slemrod, M. (1989). The Quasi-Steady State Approximation: A case study in pertubation. SIAM Rev. 31, 446-476.

Simon, P., Lafontaine, M., Delsaut, P., Morele, Y., and Nicot, T. (2000). Trace determination of urinary 3-hydroxybenzo[a]pyrene by automated column-switching high-performance liquid chromatography. J. Chromatogr. B Biomed. Sci. Appl. 748, 337-348.

Shimada, T., Inoue, K., Suzuki, Y., Kawai, T., Azuma, E., Nakajima, T., Shindo, M., Kurose, K., Sugie, A., Yamagishi, Y., Fujii-Kuriyama, Y., and Hashimoto, M. (2002). Arylhydrocarbon receptor-dependent induction of liver and lung cytochromes P450 1A1, 1A2, and 1B1 by polycyclic aromatic hydrocarbons and polychlorinated biphenyls in genetically engineered C57BL/6J mice. Carcinogenesis 23, 1199-1207.

Straif, K., Baan, R., Grosse, Y., Secretan, B., El Ghissassi, F., and Cogliano, V. (2005). Carcinogenicity of polycyclic aromatic hydrocarbons. Lancet Oncol. 6, 931-932.

Uziel, M., and Haglund, R. (1988). Persistence of benzo[a]pyrene and 7,8-dihydro-7,8-dihydroxybenzo[a] pyrene in Fischer 344 rats: time distribution of total metabolites in blood, urine and feces. Carcinogenesis 9, 233-238.

Weisstein, Eric W. "Least Squares Fitting." From MathWorld--A Wolfram Web Resource. http://mathworld.wolfram.com/LeastSquaresFitting.html (Accessed on March 2011).

Withey, J.R., Shedden, J., Law, F.C., and Abedini, S. (1993). Distribution of benzo[a]pyrene in pregnant rats following inhalation exposure and a comparison with similar data obtained with pyrene. J. Appl. Toxicol. 13, 193-202.

4.9 Captions to figures

Figure 4-1 : Chemical structures of BaP and its 3-OHBaP metabolite.

Figure 4-2 : Conceptual model of the kinetics of BaP and 3-OHBaP in rats (see Table 4-1 for the description of abbreviations).

Figure 4-3 : Comparison of model simulations (solid and dotted lines) with experimental data (symbols) of Marie et al. (2010) on the time profiles of BaP in blood and tissues of male Sprague-Dawley rats following an intravenous injection of 40 μmol/kg bw of BaP. Each symbol represents experimental means and vertical bars are standard deviations (n = 4). (A) ▲, lung; ●, adipose tissues; (B) ◆, liver; , blood; (C) ▼, skin; ■ kidney.

Figure 4-4 : Comparison of model simulations (solid and dotted lines) with experimental data (symbols) of Marie et al. (2010) on the time profiles of 3-OHBaP in blood and tissues of male Sprague-Dawley rats following an intravenous injection of 40 μmol/kg bw of BaP. Each symbol represents experimental means and vertical bars are standard deviations (n = 4). (A) ▲, lung; ●, adipose tissues; (B) ◆, liver; , blood; (C) ▼, skin; ■ kidney.

Figure 4-5 : Comparison of model simulations (solid and dotted lines) with experimental data (symbols) of Marie et al. (2010) on the cumulative excretion time courses of 3-OHBaP in urine and BaP in urine and faeces of male Sprague-Dawley rats following an intravenous injection of 40 μmol/kg bw of BaP. Each symbol represents experimental means for different groups of rats at each time point and vertical bars are standard deviations (n = 4). ■, 3-OHBaP in les feces (right axis); ●, BaP in feces; ▲, 3-OHBaP in urine (left axis).

Figure 4-6 : Comparison of model simulations (solid line) with experimental data (symbols) of Bouchard and Viau (1996) on the urinary cumulative excretion time courses of 3-OHBaP in the urine of male Sprague-Dawley rats following an intravenous injection of 40 μmol/kg bw of BaP (supposing an average animal weight of 130 g). Each symbol represents experimental means and vertical bars are standard deviations (n = 5).

Figure 4-7 : Comparison of model simulations (solid line) with experimental data (symbols) of Lee et al. (2003) on the urinary cumulative excretion time courses of 3-OHBaP in the urine of male Sprague-Dawley rats following an intraperitoneal injection of 20 mg/kg bw of BaP (supposing an average animal weight of 200 g and a constant 25 mL of urine excreted per rat per period). Each symbol represents experimental means (n = 5).

Figure 4-8 : Comparison of model simulations with the human data (symbols) of Lafontaine et al. (2004) on the urinary excretion time courses of 3-OHBaP in the urine of a worker of a carbon disk factory following an occupational exposure to BaP. The occupational scenario (gray bars) was obtained from Lafontaine et al. (2004) with two shifts of 6.75 h and 4.75 h. The worker's simulation considered an atmospheric concentration of 1514 ng/m^3 and 3028 ng/m^3, a ventilation rate of 1.20 m^3/h and an absorption fraction of 7.98%. Dashed line represents simulation based on the rat model parameter values while the solid line shows simulation with animal-to-human extrapolated parameter values.

4.10 Tables

Table 4-1 : Variables and Parameters of the Toxicokinetic Model for BaP and 3-OHBaP

Variables and parameters Symbols or abbreviations		Description
Variables	B(t)	Burden of BaP (mol) in blood as a function of time (h)
	L(t)	Burden of BaP (mol) in liver as a function of time (h)
	S(t)	Burden of BaP (mol) in skin as a function of time (h)
	K(t)	Burden of BaP (mol) in kidney as a function of time (h)
	LU(t)	Burden of BaP (mol) in lung as a function of time (h)
	AT(t)	Burden of BaP (mol) in adipose tissues as a function of time (h)
	$GI_1(t)$	Burden of BaP (mol) in the first segment of the gastrointestinal tract as a function of time (h)
	$GI_2(t)$	Burden of BaP (mol) in the second segment of the gastrointestinal tract as a function of time (h)
	U(t)	Cumulative urinary amounts of BaP as a function of time (h)
	F(t)	Cumulative fecal amounts of BaP as a function of time (h)
	OLU(t)	Burden of metabolites of BaP other than 3-OHBaP (mol) resulting from lung metabolism as a function of time (h)
	OL(t)	Burden of metabolites of BaP other than 3-OHBaP (mol) resulting from liver metabolism as a function of time (h)
	b(t)	Burden of 3-OHBaP (mol) in blood as a function of time (h)
	l(t)	Burden of 3-OHBaP (mol) in liver as a function of time (h)
	s(t)	Burden of 3-OHBaP (mol) in skin as a function of time (h)
	k(t)	Burden of 3-OHBaP (mol) in kidney as a function of time (h)
	lu(t)	Burden of 3-OHBaP (mol) in lung as a function of time (h)
	at(t)	Burden of 3-OHBaP (mol) in adipose tissues as a function of time (h)
	$gi_1(t)$	Burden of 3-OHBaP (mol) in the first segment of the gastrointestinal tract as a function of time (h)
	$gi_2(t)$	Burden of 3-OHBaP (mol) in the second segment of the gastrointestinal tract as a function of time (h)
	u(t)	Cumulative urinary amounts of 3-OHBaP as a function of time (h)
	f(t)	Cumulative fecal amounts of 3-OHBaP as a function of time (h)
	ol(t)	Burden of metabolites of 3-OHBaP (mol) resulting from liver metabolism as a function of time (h)
Parameters	f_B	Fraction of BaP in blood compared to total amounts in blood and tissues in equilibrium with blood (%)
	f_L	Fraction of BaP in liver compared to total amounts in blood and tissues in equilibrium with blood (%)
	f_S	Fraction of BaP in skin compared to total amounts in blood and tissues in equilibrium with blood (%)
	f_K	

103

Variables and parameters Symbols or abbreviations	Description
f_N	Fraction of BaP in kidney compared to total amounts in blood and tissues in equilibrium with blood (%)
K_{BL}	Fraction of BaP in non-observed organs compared to total amounts in blood and tissues in equilibrium with blood (%)
K_{BS}	Transfer rate of BaP from blood to liver (h^{-1})
K_{BK}	Transfer rate of BaP from blood to skin (h^{-1})
K_{BLU}	Transfer rate of BaP from blood to kidney (h^{-1})
K_{BAT}	Transfer rate of BaP from blood to lung (h^{-1})
K_{BN}	Transfer rate of BaP from blood to adipose tissues (h^{-1})
K_{LB}	Transfer rate of BaP from blood to non-observed tissues in equilibrium with blood (h^{-1})
K_{SB}	Transfer rate of BaP from liver to blood (h^{-1})
K_{KB}	Transfer rate of BaP from skin to blood (h^{-1})
K_{LUB}	Transfer rate of BaP from kidney to blood (h^{-1})
K_{ATB}	Transfer rate of BaP from lung to blood (h^{-1})
K_{NB}	Transfer rate of BaP from adipose tissues to blood (h^{-1})
K_{LGI}	Transfer rate of BaP from non-observed tissues to blood (h^{-1})
K_{GIGI}	Transfer rate of BaP from liver to the gastrointestinal tract (h^{-1})
K_{GIF}	Transfer rate of BaP from the first segment of the gastrointestinal tract to the second segment (h^{-1})
K_{KU}	Transfer rate of BaP from the second segment of the gastrointestinal tract to feces (h^{-1})
f_b	Transfer rate of BaP from kidney to urine (h^{-1})
f_l	Fraction of 3-OHBaP in blood compared to total amounts in blood and tissues in equilibrium with blood (%)
f_s	Fraction of 3-OHBaP in liver compared to total amounts in blood and tissues in equilibrium with blood (%)
f_n	Fraction of 3-OHBaP in skin compared to total amounts in blood and tissues in equilibrium with blood (%)
k_{bl}	Fraction of 3-OHBaP in non-observed organs compared to total amounts in blood and tissues in equilibrium with blood (%)
k_{bs}	Transfer rate of 3-OHBaP from blood to liver (h^{-1})
k_{bk}	Transfer rate of 3-OHBaP from blood to skin (h^{-1})
k_{blu}	Transfer rate of 3-OHBaP from blood to kidney (h^{-1})
k_{bat}	Transfer rate of 3-OHBaP from blood to lung (h^{-1})
k_{lb}	Transfer rate of 3-OHBaP from blood to adipose tissues (h^{-1})
k_{sb}	Transfer rate of 3-OHBaP from liver to blood (h^{-1})
k_{kb}	Transfer rate of 3-OHBaP from skin to blood (h^{-1})
k_{lub}	Transfer rate of 3-OHBaP from kidney to blood (h^{-1})
k_{atb}	Transfer rate of 3-OHBaP from lung to blood (h^{-1})
k_{lgi}	Transfer rate of 3-OHBaP from adipose tissues to blood (h^{-1})
k_{gigi}	Transfer rate of 3-OHBaP from liver to the gastrointestinal tract (h^{-1})
k_{gil}	Transfer rate of 3-OHBaP from the first segment of the gastrointestinal tract to the second segment (h^{-1})

Variables and parameters Symbols or abbreviations	Description
k_{gif}	Transfer rate of 3-OHBaP from the first segment of the gastrointestinal tract back to liver (h^{-1})
k_{ku}	Transfer rate of 3-OHBaP from the second segment of the gastrointestinal tract to feces (h^{-1})
K_{LUlu}	Transfer rate of 3-OHBaP from kidney to urine (h^{-1})
K_{LI}	
K_{LUOLU}	Metabolism rate of BaP into 3-OHBaP in lung (h^{-1})
K_{LOL}	Metabolism rate of BaP into 3-OHBaP in liver (h^{-1})
k_{lol}	Metabolism rate of BaP into metabolites other than 3-OHBaP in lung (h^{-1})
	Metabolism rate of BaP into metabolites other than 3-OHBaP in liver (h^{-1})
	Metabolism rate of 3-OHBaP into other metabolites in liver (h^{-1})

Table 4-2 : Parametric Values of the Toxicokinetic Model for BaP and 3-OHBaP

Parameters		Values
Fractional tissue partition with	f_B	9.14×10^{-1}
reference total amounts in blood	f_L	12.80
and tissues in equilibrium with	f_S	6.27
blood (%)[a]	f_K	9.35×10^{-1}
	f_N	79.08
	f_b	2.67
	f_l	5.07
	f_s	4.12
	f_n	88.14
Transfer rates (h^{-1})		
	K_{BLU}	4.28
	K_{BAT}	0.952
	K_{LUB}	3.19×10^{-3}
	K_{ATB}	2.55×10^{-2}
	K_{LGI}	5.89×10^{-3}
	K_{GIGI}	4.96×10^{-2}
	K_{GIF}	1.25×10^{-1}
	K_{KU}	3.68×10^{-1}
	k_{bk}	1.50
	k_{blu}	1.63×10^{-2}
	k_{bat}	1.47×10^{-1}
	k_{kb}	1.44×10^{-1}
	k_{lub}	4.58
	k_{atb}	9.76×10^{-2}
	k_{lgi}	3.81
	k_{gigi}	7.39×10^{-2}
	k_{gil}	1.77×10^{-2}
	k_{gif}	1.25×10^{-1}
	k_{ku}	9.11×10^{-3}
	K_{LUlu}	3.91×10^{-3}
	K_{LI}	2.09×10^{-1}
	K_{LUOLU}	1.66×10^{-2}
	K_{LOL}	8.68×10^{-1}
	k_{lol}	1.37

[a]These fractions are defined as follows: $f_X = \dfrac{BaP\ in\ compartment\ X}{BaP\ in\ all\ tissues\ in\ equilibrium\ with\ blood}$.

Table 4-3 : Elimination Half-lives of BaP and 3-OHBaP in Blood and Tissues, as Established from Model Parameter Values

Tissue	Elimination half-life (h)					
	BaP			3-OHBaP		
	Parameter	Value	χ^{2a}	Parameter	Value	χ^{2a}
Blood	$t_{\frac{1}{2}BAT}$	0.728		$t_{\frac{1}{2}bat}$	4.76	
	$t_{\frac{1}{2}BLU}$	0.162	0.456	$t_{\frac{1}{2}blu}$	42.7	0.014
				$t_{\frac{1}{2}bk}$	0.463	
Liver	$t_{\frac{1}{2}LI}$	3.32		$t_{\frac{1}{2}lgi}$	0.182	
	$t_{\frac{1}{2}LGI}$	118	15.8	$t_{\frac{1}{2}lol}$	0.505	0.809
	$t_{\frac{1}{2}LOL}$	0.799				
Kidney	$t_{\frac{1}{2}KU}$	1.89	0.410	$t_{\frac{1}{2}ku}$	76.1	0.063
Skin	-	-	8.15	-	-	0.511
Faeces	$t_{\frac{1}{2}GIGI}$	14.0	0.324	$t_{\frac{1}{2}gigi}$	9.38	19.4
	$t_{\frac{1}{2}GIF}$	5.55		$t_{\frac{1}{2}gif}$	5.55	
Urine	-	-	-	-	-	0.139
Adipose tissues	$t_{\frac{1}{2}AT}$	27.2	5.88	$t_{\frac{1}{2}at}$	7.10	0.132
Lung	$t_{\frac{1}{2}LUB}$	217		$t_{\frac{1}{2}lub}$	0.151	
	$t_{\frac{1}{2}LUlu}$	177	0.304			0.122
	$t_{\frac{1}{2}LUOLU}$	41.9				

[a] χ^2 is the coefficient of goodness of fit defined by:

$$\chi^2 = \sum_{\forall i} \frac{\left(y_i^{exp} - y_i^{theo}\right)^2}{\sigma^2},$$ where the exponent [exp] refers to experimental data, [theo] is related to

model prediction of data points at time i and σ represents the standard deviation from the corresponding data set.

4.11 Figures

Figure 4-1: Chemical structures of BaP and its 3-OHBaP metabolite.

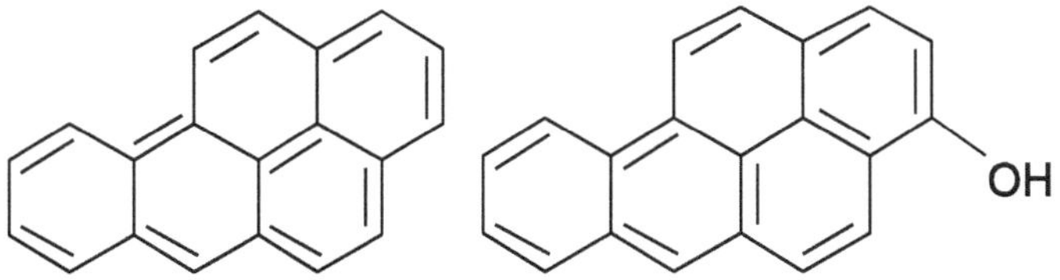

Benzo(a)pyrene   **3-Hydroxybenzo(a)pyrene**

(BaP)   **(3-OHBaP)**

Figure 4-2: Conceptual model of the kinetics of BaP and 3-OHBaP in rats (see Table 4-1 for the description of abbreviations).

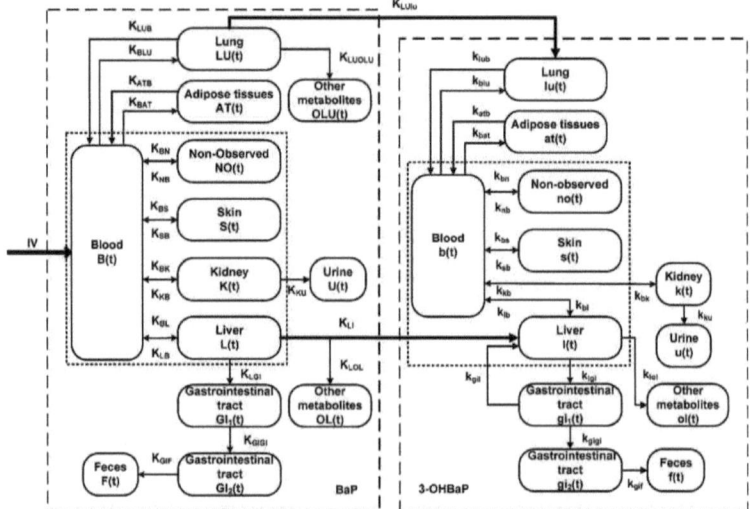

Figure 4-3: Comparison of model simulations (solid and dotted lines) with experimental data (symbols) of Marie et al. (2010) on the time profiles of BaP in blood and tissues of male Sprague-Dawley rats following an intravenous injection of 40 μmol/kg bw of BaP.

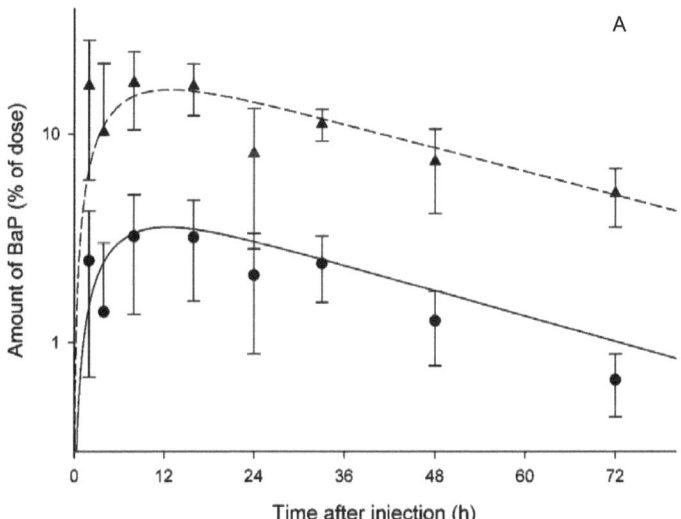

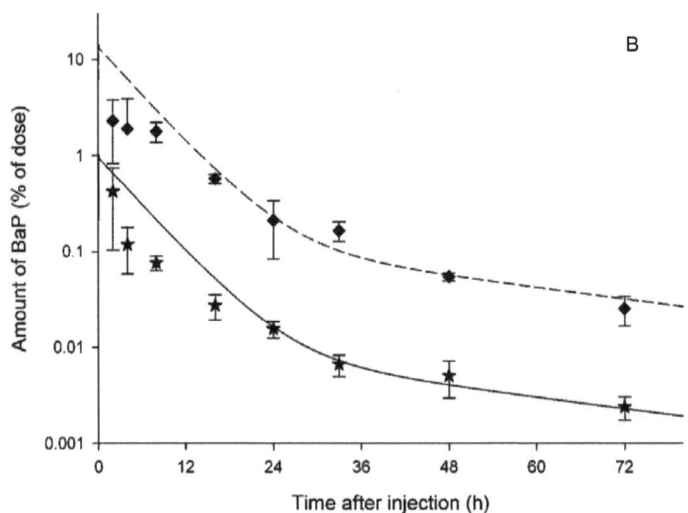

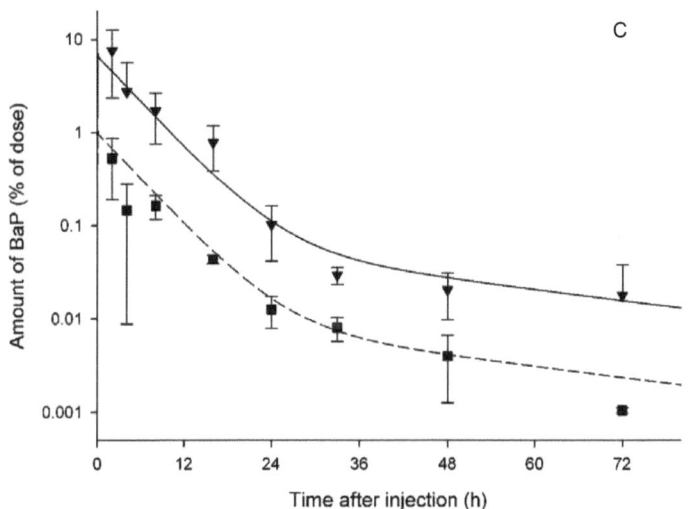

Figure 4-4: Comparison of model simulations (solid and dotted lines) with experimental data (symbols) of Marie et al. (2010) on the time profiles of 3-OHBaP in blood and tissues of male Sprague-Dawley rats following an intravenous injection of 40 μmol/kg bw of BaP.

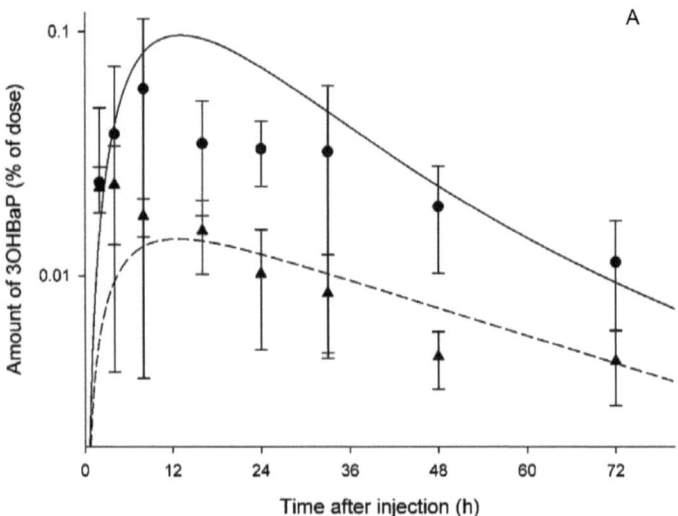

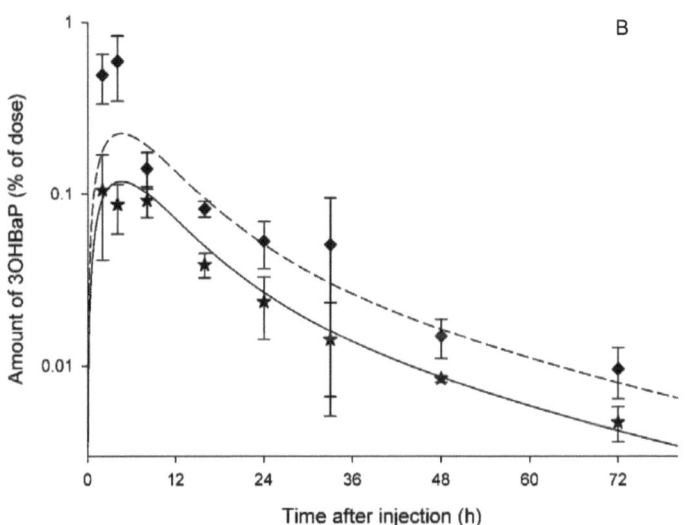

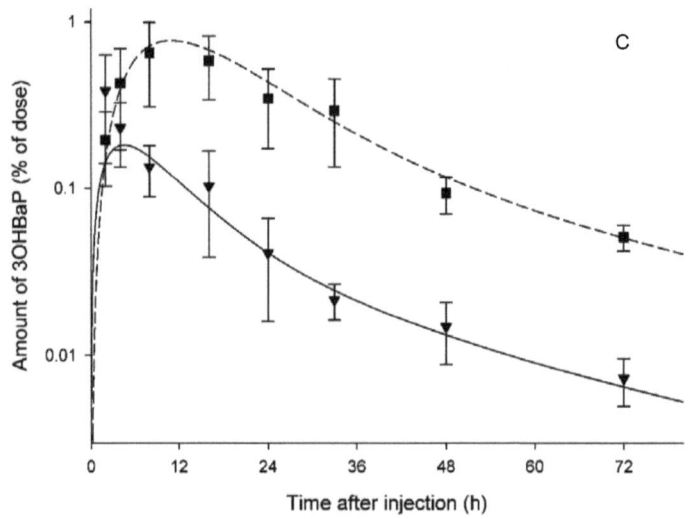

Figure 4-5: Comparison of model simulations (solid and dotted lines) with experimental data (symbols) of Marie et al. (2010) on the cumulative excretion time courses of 3-OHBaP in urine and BaP in urine and faeces of male Sprague-Dawley rats following an intravenous injection of 40 μmol/kg bw of BaP.

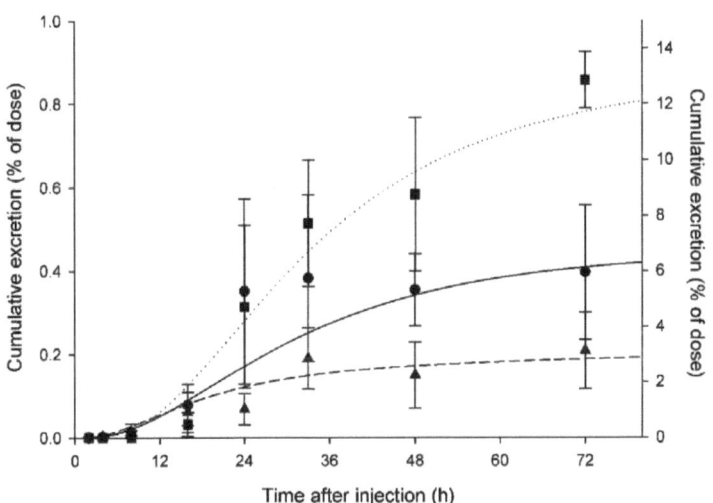

Figure 4-6: Comparison of model simulations (solid line) with experimental data (symbols) of Bouchard and Viau (1996) on the urinary cumulative excretion time courses of 3-OHBaP in the urine of male Sprague-Dawley rats following an intravenous injection of 40 µmol/kg bw of BaP (supposing an average animal weight of 130 g).

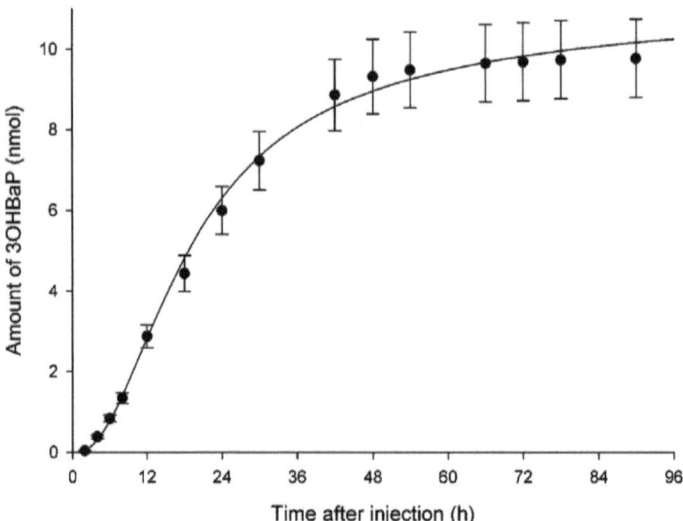

Figure 4-7: Comparison of model simulations (solid line) with experimental data (symbols) of Lee et al. (2003) on the urinary cumulative excretion time courses of 3-OHBaP in the urine of male Sprague-Dawley rats following an intraperitoneal injection of 20 mg/kg bw of BaP (supposing an average animal weight of 200 g and a constant 25 mL of urine excreted per rat per period).

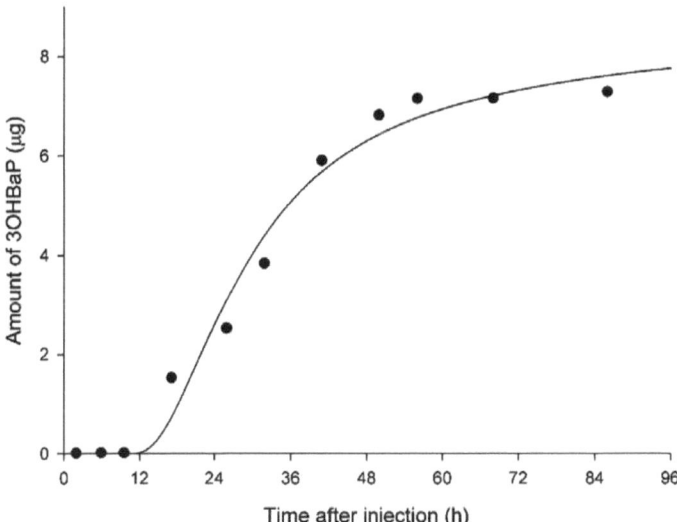

Figure 4-8: Comparison of model simulations with the human data (symbols) of Lafontaine et al. (2004) on the urinary excretion time courses of 3-OHBaP in the urine of a worker of a carbon disk factory following an occupational exposure to BaP.

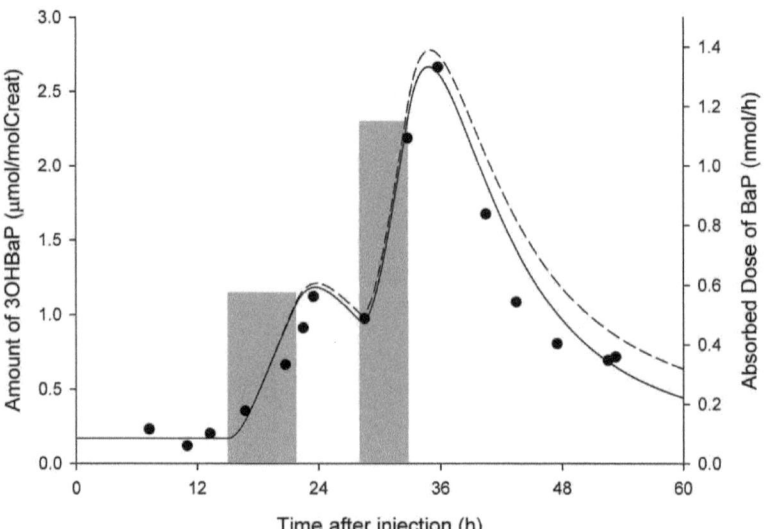

5 Deuxième article : Understanding the linked kinetics of benzo(a)pyrene and 3-hydroxybenzo(a)pyrene biomarker of exposure using physiologically-based pharmacokinetic modelling in rats

Roberto Heredia-Ortiz and Michèle Bouchard

Heredia-Ortiz, R. et Bouchard, M. Understanding the linked kinetics of benzo(a)pyrene and 3-hydroxybenzo(a)pyrene biomarker of exposure using physiologically-based pharmacokinetic modelling in rats. J Pharmacokinet Pharmacodyn. 2013 Dec; 40(6): 669-82. doi: 10.1007/s10928-013-9338-9. Epub 2013 Oct 29.

Understanding the linked kinetics of benzo(a)pyrene and 3-hydroxybenzo(a)pyrene biomarker of exposure using physiologically-based pharmacokinetic modelling in rats

Roberto Heredia-Ortiz[*] and Michèle Bouchard[*,†]

[*]Department of Environmental and Occupational Health, Chair of Toxicological Risk Assessment and Management, and Public Health Research Institute (IRSPUM), Faculty of Medicine, University of Montreal, Montreal, Canada

[†] Corresponding author:

Michèle Bouchard, Ph.D.
Associate professor
Department of Environmental and Occupational Health
University of Montreal
P.O Box 6128, Main Station, Montreal (QC)
H3C 3J7, CANADA
Telephone number: (514) 343-6111 ext 1640
Fax number: (514) 343-2200

124

5.1 Abstract

3-Hydroxybenzo(a)pyrene (3-OHBaP) in urine has been proposed as a biomarker of occupational exposure to polycyclic aromatic hydrocarbons (PAHs). However, to reconstruct exposure doses in workers from biomarker measurements, a thorough knowledge of the kinetics of the benzo(a)pyrene (BaP) and 3-OHBaP given different routes of exposure is needed. A rat physiologically-based pharmacokinetic (PBPK) model of BaP and 3-OHBaP was built. Organs (tissues) represented as compartments were based on in vivo experimental data in rats. Tissue: blood partition coefficients, permeability coefficients, metabolism rates, excretion parameters, and absorption fractions and rates for different routes-of-entry were obtained directly from published *in vivo* time courses of BaP and 3-OHBaP in blood, various tissues and excreta of rats. The latter parameter values were best-fitted by least square procedures and Monte Carlo simulations. Sensitivity analyses were then carried out to ensure the stability of the model and the key parameters driving the overall modeled kinetics. This modeling pointed out critical determinants of the kinetics: 1) hepatic metabolism of BaP and 3-OHBaP elimination rate as the most sensitive parameters; 2) the strong partition of BaP in lungs compared to other tissues, followed by adipose tissues and liver; 3) the strong partition of 3-OHBaP in kidneys; 4) diffusion-limited tissue transfers of BaP in lungs and 3-OHBaP in lungs, adipose tissues and kidneys; 5) significant entero-hepatic recycling of 3-OHBaP. Very good fits to various sets of experimental data in rats from four different routes-of-entry (intravenous, oral, dermal and inhalation) were obtained with the model.

Keywords: PBPK modeling; benzo(a)pyrene; 3-hydroxybenzo(a)pyrene, biomarkers of exposure

5.2 Introduction

Polycyclic aromatic hydrocarbons (PAHs) are common occupational and environmental contaminants. Benzo(a)pyrene (BaP) is one of the most potent PAH, classified as a known human carcinogen [1]. Recently, measurement of its 3-hydroxybenzo[a]pyrene (3-OHBaP) metabolite in urine of workers has been proposed to assess occupational exposure to BaP and by extension to PAHs.

To allow a proper interpretation of biomonitoring data and reconstruct dose from biomonitoring data, kinetic modeling is increasingly used [2,3]. A physiologically-based pharmacokinetic (PBPK) model has been proposed by Moir et al. and Clowell et al. [4,5] to simulate the kinetics of BaP but such model did not aim at establishing the relation between BaP kinetics and that of its metabolites, such as 3-OHBaP. In the model of Moir et al. [4], parameter values such as tissue: blood partition coefficients were determined from *in vitro* data, but it is also possible to develop a method for the *in vivo* determination of these parameters from the time-courses of the compound of interest in blood and tissues of animals. Since the tissue: blood partition and tissue permeability coefficients can be found to be proportional to the typical octanol-water partition coefficients [6,7], one expects those coefficients to remain almost constant between animals and humans [8]. The same values determined *in vivo* in animals could then be used directly in a PBPK model extrapolated to humans.

The general objective of the present study was to develop a PBPK model to describe the kinetics of BaP and its metabolite 3-OHBaP in rats for different exposure routes and temporal scenarios, based on available *in vivo* time course data, to set the appropriate basis for a human-extrapolated model.

5.3 Materials and Methods

5.3.1 Conceptual representation of the model

A PBPK model representing the kinetics of BaP and its metabolite 3-OHBaP was developed. Organs (tissues) represented as compartments are based on in vivo experimental data and toxicokinetic modeling performed in rats by our team [9,10]. The conceptual representation of the model is shown in Figure 5-1. The kinetics of BaP and 3-OHBaP were simulated for four different routes of exposure: inhalation, intravenous, oral, and skin.

For each organ, it was determined whether the transfer was limited by tissue perfusion or diffusion (permeability), hence if a tissue: blood equilibrium was reached instantaneously or if transfer to the tissue was limited by diffusion (permeability) of the molecule through the cell membranes. For each of these processes, different differential equations must be considered. This step precedes optimization of model parameters. To define which organ exhibited a diffusion-limited transfer, we used a novel data-based method where peak concentration was determined mathematically for each organ simply by assuming that the kinetics were only limited by the perfusion for the entire system (each maximum tissue concentration being proportional to the tissue volume). Then, when the value obtained was in agreement with the experimental data of Marie et al. [9], we considered that the tissue distribution was limited by the tissue perfusion, otherwise a cell/tissue distribution limited by diffusion was considered [11]. More specifically, tissue perfusion-limited processes were described only by two free parameters per compartment: the input and the output rates. The input rate is not really a "free" parameter given that it is determined by Q_{tissue} (the blood flow rate to a given tissue). The output rate is determined by the ratio $Q_{tissue}/(P_{tissue}V_{tissue})$, hence by the tissue volume/weight V_{tissue} and the tissue partition coefficient P_{tissue}. Therefore, the output rate is fixed by the tissue partition coefficient value and determines the elimination slope of the tissue time-course curve. In other words, by fixing this value to best fit the observed elimination slope of the time course curve, we are automatically fixing the maximum value and the corresponding time-to-maximum concentration. If the tissue kinetics is truly perfusion-limited, these two parameters are sufficient to describe the complete shape of the tissue-time course. Otherwise, it means that other mechanisms intervene, namely, tissue diffusion kinetics (e.g. good simulation of the slope but underestimation of tissue levels of BaP or 3-OHBaP, good estimation of maximum amounts of BaP or 3-OHBaP but underestimation of elimination rate, *etc*.).

In the model, all compartments were considered to be in equilibrium with venous blood at any time and no organ was considered to be in complete state of kinetic equilibrium. In addition, the kinetics of BaP and 3-OHBaP in the whole blood were modeled; hence, the model developed does not consider sub-compartments and thus distinct dynamics between red blood cells and plasma. Regarding the hepatic kinetics, we have considered a combined source of blood flow from the arterial and hepatic portal vein for BaP while we have taken into account an entero-hepatic recycling for 3-OHBaP. Based on the toxicokinetic modeling reported in Heredia et al. [10], it was also considered that the lung and skin biotransformation was negligible compared to hepatic metabolism. We have further assumed that the oral absorption rate was a combination of uptake through the epithelium of gastrointestinal tract and the rate of transfer from blood of the hepatic portal vein to the liver. This rate thus represents the total effective absorption rate of orally administered BaP by the different parts of the GI tract (including the stomach and small intestine) and delivery to the liver (by hepatic portal vein). The oral absorption fraction corresponds to the fraction transferred from the GI tract to the liver (absorption) versus the unabsorbed fraction transferred to the faeces (faecal elimination).

The proposed PBPK model considers linear kinetics and is thus represented by first order differential equations. Therefore, all differential equations used in model development are

128

linear combinations leading to infinite domains of validity. However, our model has only been validated from a range of doses of 1.7 nmol to 9.9 μmol.

5.3.2 Kinetic parameters

All values of physiological parameters such as organ weights, blood volume and tissue blood flow rates (% of cardiac output) used to model the compartmental kinetics correspond to values commonly used in rats (as prescribed in Brown et al. [12]). Tissue: blood partition coefficients, tissue permeability coefficients, metabolism rates, excretion parameters, and absorption rates and fractions for different routes of entry (inhalation, dermal, oral) were obtained directly from *in vivo* rat data available in the literature. More specifically, the experimental data shown in Table 5-1 were used to determine unknown parameter values. Briefly, once the volumes and physiological blood flow rates were determined for each organ/tissue, the remaining parameters to determine were the partition coefficients in the case of organs with perfusion-limited transfers and permeability coefficients for tissues with diffusion-limited processes, as well as the rate of hepatic metabolism.

A set of coupled Monte Carlo algorithms have been generated to calculate all possible values for these unknown parameters (see, e.g., [13]). Matlab software was used for these iterations. The details of this process are shown schematically in Figure 5-2. This step is a central novel aspect of this project. Similar attempts have been reported [14-21] but most researchers calculate very specific *in vivo* parameters, while relying on *in vitro* data for most values of partition and permeability coefficients as well as metabolism constants in their models.

The main statistical criterion used to test the quality of fits of the differential equations to the experimental data was the Pearson χ^2, and the corresponding (1-p)% value [22], which is a commonly used test to determine the correspondence of a general function to a set of

Gaussian-distributed points. The experimental data were considered log-normally distributed (Galton distribution), therefore easily transformed to normal distributions.

5.3.3 Model simulations

Once all the parameters of the model were fixed, using Monte Carlo runs, simulations were carried out by numerically solving the system of differential equations representing the kinetics of BaP, on the one hand, and 3-OHBaP, on the other hand. All simulations were performed using Matlab.

5.3.4 Sensitivity analysis

Sensitivity analyses were performed to verify the stability of the model. There are several tests to measure the sensitivity of a particular parameter on a given function. In general, sensitivity analysis is a mathematical procedure in which we verify how the value of a dependent function is modified as we vary the default value of a given parameter (one independent degree of freedom), while keeping the rest of the parameters unchanged.

In addition, another type of sensitivity analysis was performed to test the stability of the model for different routes of exposure and synergistic effects between parameters. We simulated the time profile of 3-OHBaP in urine following a combined exposure to 40 μmol/kg BaP by ingestion, dermal contact, inhalation and intravenous injection. We performed a Monte Carlo analysis in these conditions, in which all model parameter values were varied stochastically. Then, we estimated the difference between the profiles resulting from the overall parameter changes and those obtained with the default parameter values (initially optimized). Of all these calculations (more than 1000 runs), 91.7% ± 1.63% of the simulations showed a variation lower than 10% compared to simulations with initially optimized parameter conditions.

5.3.5 Model validation

Finally, the model ability to simulate *in vivo* time course data other than those used to determine model parameters was verified. These data are presented in Table 5-2.

5.4 Results

The PBPK model obtained proved to be very suitable for reproducing data from Marie et al. [9] on the time profiles of BaP and 3-OHBaP in various biological matrices of rats exposed intravenously. Numerical simulations are presented in Figures 5-3 and 5-4. All values of (1-p)% for each curve showed good adjustments, well above 90%. Tissue: blood partition coefficients, elimination rates, tissue permeability coefficients and metabolism constants derived from these data are shown in Table 5-3. Sensitivity calculated for all parameters is listed in Table 5-3 as well. The most sensitive parameters were found to be the metabolic constant and urinary elimination rate.

Tissue: blood partition coefficients for BaP are very different between tissues. Lung: arterial blood partition coefficient value is several orders of magnitude higher than those of other tissues, followed by the adipose tissues and the liver. On the other hand, partition coefficient values of 3-OHBaP do not differ much from one tissue to the other, compared to the parent compound. The kidney is the organ which concentrated 3-OHBaP the most, with a tissue: blood partition coefficient about one order of magnitude higher than those of other organs. Tissue distribution (hence the different tissue: blood partition coefficients) is very different between BaP and 3-OHBaP, with the exception of the skin, which concentrates in the same manner both molecules.

Tissue distribution limited by diffusion, as represented by tissue permeability coefficients, had to be considered for BaP only in lungs and for 3-OHBaP in lungs, adipose tissues and kidneys. For BaP, the lungs exhibited the highest tissue permeability coefficient while for 3-OHBaP it was the kidney that showed the highest value. Sensitivity analysis values presented in Table 5-3 show that the permeability coefficients are not the key parameters determining the overall time courses of BaP or 3-OHBaP in the various tissues. For instance, smaller permeability coefficients would only decrease concentrations at all times proportionally in

132

organs limited by diffusion without significantly altering the curves of the other compartments. .

The metabolic constants in the model only determine the ratio between the maximum velocity rate and the Michaelis affinity constant (V_{max}/K_m) of each molecule as our model represents only the first linear approximation of the Michaelis-Menten mechanism, but this is in line with the observed lack of saturation process in liver metabolism over the studied exposure dose range (up to 9 µmol dose). Results presented in Table 5-3 show a very rapid biotransformation of BaP into its various metabolites. Since our model only tracks the 3-OHBaP metabolite, we assumed that it represents a fraction of total metabolites produced from BaP metabolism as follows:

$$\frac{V_{max}^{(eff)}}{K_m^{(eff)}} = \sum_{\forall i} \frac{V_{max}^{(i)}}{K_m^{(i)}}. \tag{5-1}$$

$$\text{fraction}_{3-OHBaP} = \frac{V_{max}^{(3-OHBaP)}}{K_m^{(3-OHBaP)}} \left(\sum_{\forall i} \frac{V_{max}^{(i)}}{K_m^{(i)}} \right)^{-1}. \tag{5-2}$$

We have also considered a rapid biotransformation of 3-OHBaP. According to the percentage of sensitivity reported in Table 5-3, it is clear that the metabolic constants of BaP and 3-OHBaP are the parameters that influence the most the urinary excretion kinetics, because even a very small variation in these parameters can cause the simulation of the expected urinary excretion to change by at least 10%.

In addition to the importance of the metabolic constants, we found that the elimination rates of 3-OHBaP largely determine the urinary excretion kinetics. The fastest 3-OHBaP transfer rate is the transfer from the liver to the bile (klgi more than two orders of magnitude higher), which leads directly to high levels of 3-OHBaP in faeces. The values of the other excretion parameters are similar, the K_{kbr} value for the 3-OHBaP transferred from kidneys to the

133

bladder being of the same order of magnitude as the rate of elimination of 3-OHBaP from the bladder to urine K_{bu} and the faecal excretion K_{gif}.

Once the PBPK model was built and the sensitivity of the parameters was established, we tested the ability of the model to reproduce accurately different independent sets of experimental data on the internal kinetics. In particular, the model was used to simulate other intravenous data available in rats [23,24]. Bouchard and Viau [23] exposed Sprague-Dawley rats (n = 4) intravenously with the same dose that has generated the experimental data used for the construction of our model. Data on the excretion time course of 3-OHBaP in urine were provided in this study over a 90-h period following intravenous injection of 40 μmol of BaP /kg body weight (bw). Similarly, Cao et al. [24] studied the excretion profile of 3-OHBaP in the urine of Sprague-Dawley rats over a 37-h period following intravenous injection of 1 mg BaP/kg bw. Figure 5-5 shows the comparison of simulated and observed kinetics without any modification of model parameters. The model adequately reproduced the urinary profiles with a (1-p)% value on the corresponding χ^2 statistics greater than 99%.

The model was also used to simulate the data available in rats exposed by inhalation or intratracheal instillation (Figure 5-6). Using the experimental data presented by Weyand and Bevan [25], we were able to adjust the respiratory parameters (see Table 5-3). For BaP, a blood: air partition coefficient relatively close to one was obtained to represent exhalation. Moreover, given the low concentrations of 3-OHBaP in the lungs, we did not include expiration of 3-OHBaP metabolite in the modeling. Our PBPK model was able to reproduce the time profile of BaP in blood reported by Weyand and Bevan in Sprague-Dawley rats (n = 3) exposed to 1 mg of BaP /kg bw by intratracheal instillation. The model was also able to simulate adequately another independent set of data from Ramesh et al. [26] on the time profile of BaP in blood of male Fisher rats (n=5) over a 4-h period following a 4-h inhalation of 100 mg/m^3 of BaP.

Furthermore, the kinetics of BaP and 3-OHBaP after dermal exposure were also simulated. The urinary excretion profile of 3-OHBaP presented by Payan et al. [27] was used to derive dermal absorption parameters and the ability of the model to simulate another independent set of cutaneous data [28] was then tested. Payan et al. determined the temporal profile of 3-OHBaP in urine after a dermal application of 70 mg of BaP in Sprague-Dawley rats while Jongeneelen et al. [28] have documented the time profile in male Wistar rats after dermal application of 50 µmol/kg body weight of BaP. As shown in Figure 5-7, the model adequately reproduced the experimental data used to fit skin parameters (see Table 5-3), but was also able to reproduce the independent experimental data of Jongeneelen et al. [28].

Finally, to establish the oral absorption rate (see Table 5-3), the experimental data of Cao et al. [24] on the time profile of BaP in blood of Sprague-Dawley rats after an exposure to 20 mg of BaP/kg bw were used. Our single compartment modeling of the GI tract was found to adequately simulate oral time profiles since, as described in Crowell et al. [5], the oral absorption rate is expected to have a first order behavior and can therefore be modeled as a global effective mechanism instead of several small steps. It was then verified that the model adequately simulated the corresponding 3-OHBaP data reported by the same authors. Figure 8 shows the respective PBPK simulations without any adjustment of model parameter values. The model adequately reproduced the time courses by different routes of exposure (Figures 5-6, 5-7 and 5-8) with a χ^2 corresponding to a (1-p)% value above 90%.

Other observed time course results, listed in Table 5-1, have also been simulated. In all the cases, our PBPK model was able to accurately reproduce the observed concentrations of BaP or 3-OHBaP in all of the studied tissues and fluids, within an order of magnitude. However, some of the experiments, on BaP only (particularly, Moir et al. [29]), have shown a steeper elimination slope than the one predicted by our model. Furthermore, differences observed in the studies mentioned can be a result of various factors such as: very different methodologies to measure the parent compound and to isolate the metabolites, possible differences in

135

accounted free and conjugated molecules, different species of rats, very short periods of sampling and a larger range of doses administered.

5.5 Discussion

To our knowledge, no PBPK model has been published to describe and explain the overall linked kinetics between BaP and its 3-OHBaP biomarker of exposure. Ciffroy et al. [30] presented an integrated modeling approach for predicting internal tissue concentrations of BaP from water intake by coupling a multimedia environmental model and a PBPK model. Unfortunately, they did not present any details on the PBPK model developed that can be used for other purposes. Actually, the model presented in their paper was an adaptation of a more generic PBPK mainly based on a related PAH, pyrene. Recently, Crowell et al. [5] published a model similar to the one developed at the same time by our team for BaP, but the model focussed only on the parent compound. Similarly, Haddad et al. [31] and Jongeneelen and Berg [32] proposed a PBPK model to simulate exposure to pyrene and its 1-hydroxypyrene metabolite, which, even though could be a good starting point for the kinetics of BaP and 3-OHBaP (e.g the relevant organs and their diffusion-limited features), does not directly represent the intricate behavior observed for BaP and 3-OHBaP.

From model simulation of the time course of BaP and 3-OHBaP in blood and tissues in dynamic equilibrium with blood, a bi-exponential elimination behaviour was observed. Each exponential trend reflects different biological processes. The steepest slope is modeled to be driven by the rapid biotransformation of BaP following exposure to BaP. The second phase is modeled to be associated with the slow release of BaP from adipose tissues that act as storage tissues. There is good evidence of binding of BaP to lipoproteins and the influence of plasma lipoproteins and albumin as carriers of BaP in rats [33]. Later, Sugihara and James [34] suggested, in an *in vitro* bovine study, that 3-OHBaP metabolite could also bind covalently to blood proteins. Our PBPK model was not able to distinguish between lipid storage and protein binding of BaP and 3-OHBaP since they are mathematically equivalent representations. In order to account for the respective contribution of protein binding and adipose tissue storage, it would require experimentally tracking the free and the bound

137

components of BaP and 3-OHBaP, which is unfortunately unavailable in the current literature.

Concerning the routes-of-entry, Sagredo et al. [35] suggested that the distribution and tissue levels of BaP and BaP-protein adducts are clearly dependent on the route of exposure. Very few inhalation studies in animals are available; the study of Ramesh et al. [26] used in the current modeling and that of Wang et al. [36], who characterized BaP and its hydroxy metabolites in the urine of 24 rats exposed to asphalt fumes, are about the only available studies. Contrary to inhalation and oral exposure, one would expect a slower dermal absorption of BaP. However, our modeling rather showed that the slow release of BaP from adipose tissues and lungs (possibly lipid components of the lung) was the rate-limiting step in the kinetics and driving the overall observed time profiles of BaP in blood.

Modeling of the various available data appears to indicate that liver metabolism dominated the overall metabolism of BaP and even 3-OHBaP metabolite. In other words, with a model considering negligible biotransformation in organs other than the liver, we were able to reproduce the blood and tissue time courses for different routes-of-entry over time ranges relevant for biomonitoring (hours). However, there exists some evidence of BaP biotransformation in different organs [37]. Most *in vitro* studies refer to a significant BaP biotransformation capacity of the liver but also the lungs. Nevertheless, *in vivo*, it is expected that the lungs will play an important role in the kinetics at very short times following an inhalation exposure (minutes). From a biomonitoring standpoint using BaP metabolites, pertinent time scales are in the hour range rather than minute range, hence when the liver becomes the main BaP biotransformation contributor. Weyand and Bevan [25] reported that 59.6% of an intratracheal ^{3}H-labelled BaP dose in Sprague-Dawley rats were recovered as radioactivity in lungs five minutes post-exposure and this percentage fell to 15.4% one hour post-dosing, while liver radioactivity represented 15.4 and 15.8% of dose five minutes and one hour post-dosing, respectively. Molliere et al. [38] also showed that perfused lungs from

138

rats released free and conjugated metabolites of BaP to the perfusate; however, when the liver was incorporated to the perfusion circuit, the metabolites found were reduced to less than 20% of the concentrations found in the absence of the liver, which shows the major contribution of the metabolism from the liver over the lungs. Furthermore, Molliere et al. [39] concluded that liver metabolism was more rapid than that of lungs, even though total metabolite formation in the lungs after two hours was comparable with that of the liver. Finally, with regard to the metabolic constant values found, the kinetic model built is very dependent on those values. The metabolic rate constants of BaP into total metabolites (as well as the fraction of 3-OHBaP produced) and of 3-OHBaP to other metabolites are among the most sensitive parameters in the system (Table 5-3). Since there is evidence that there may be wide variations in those values (e.g. 9% in rodents and 65.9% in humans [39,52,53]), this intra-individual and inter-species sensitivity must be taken into account whenever this kind of model is employed. Variations in metabolic constant values were found to have an impact mainly on the percentage of BaP dose recovered as 3-OHBaP in urine but had limited effect on the shape of the time course curve of this biomarker of exposure.

In a dermal study, DiGiovanni et al. [40] showed that cultures of mouse epidermal cells effectively metabolized ^{3}H-BaP with trans-7,8-dihydro-7,8-dihydroxybenzo(a)pyrene and 3-OHBaP as the major organic solvent-extractable metabolites found in the intracellular cultures; the water-soluble metabolites found in the extracellular medium were conjugated to glucuronic acid (primarily 3-OHBaP and several quinones). However, our model was able to suitably represent the dermal exposure and skin irrigation without any form of metabolism in the skin.

Chipman et al. [41] provided evidence supporting the importance of the enterohepatic recycling of 3-OHBaP in rabbits as a result of observations of 3-OHBaP in the bile. Van Schooten et al. [42] later measured 3-OHBaP in blood, faeces or urine of rats exposed to contaminated soil containing BaP and found non linear pharmacokinetic parameters in their

139

study, which also lead them to suggest the importance of enterohepatic recycling in the kinetics. Our model corroborated their findings by including an enterohepatic recycling of 3-OHBaP, which proved to be very important in the kinetics of this BaP metabolite.

The atypical time course of 3-OHBaP in kidneys and urine was expected to result from the difference between the relatively rapid uptake of 3-OHBaP in the kidneys and its slower release; the underlying mechanism may be a differential rate of transfer between the apical and basolateral transporters of kidney tubules. Our model also assumed a simple transition time in the bladder (no metabolism), which allowed to reproduce the slight temporal shift in 3-OHBaP time courses observed in urine compared to kidneys. This may result from a possible interaction of 3-OHBaP in the bladder. Using rat and human cell cultures, Moore et al. [43] found evidence that BaP can be metabolized in the bladder of both species, forming similar metabolites but in slightly different proportions. They also reported that BaP metabolites could covalently bind to DNA of urothelial tissues. Finally, 3-OHBaP in kidneys exhibited a slow elimination (with respect to its blood perfusion input), which was reflected by an observed delay in time-to-peak levels of a few hours compared to other tissues.

The model developed in rats using an *in vivo* determination of parameter values, not only provided a better understanding of the relationship between BaP and its biomarker of exposure 3-OHBaP but also sets the basis for the establishment of a human PBPK model, including most sensitive parameters to consider. In particular, results of sensitivity analysis showed that we must be very careful when applying the proposed PBPK model to subjects that may have a metabolism different from that of male Sprague-Dawley rats. Tissue: blood partition coefficients determined from *in vivo* time profiles, and which are considered similar between rats and humans, will also be valuable parameters values in the extrapolated model. The human model will serve to reconstruct absorbed doses of BaP in workers for which repeated biological measurements are collected. More importantly, as done for other compounds [2,44,3], it will also allow estimating biological reference values (BRV), hence

140

urinary biomarkers levels of 3-OHBaP corresponding to an available exposure dose limits for BaP. It will then be possible to directly compare urinary excretion values measured in workers with the proposed BRV.

5.6 Acknowledgments

This work was supported by the Agence nationale de sécurité sanitaire de l'alimentation, de l'environnement et du travail (ANSES) [Grant number 2011-CRD-13] and the Chair in Toxicological Risk Assessment and Management of the University of Montreal.

5.7 Appendix

The following set of linear equations was used to mathematically represent the system described in Figure 5-1. As usual, C stands for concentrations (μmol/mL), A for amounts (μmol), Q for regional blood flow, V for volumes, P for tissue: blood partition coefficients, PA for permeability-area coefficients, f for fractions (e.g. fraction of blood in tissues), K_P for skin permeability coefficient, K for rates, S surface of exposure, V_{MAX} total maximum rate of metabolism and K_m for Michaelis affinity constant. Finally, tissues and fluids are labeled (subscripts) as follows: LU for lungs, AT for adipose tissues, V for venous blood, A for arterial blood, S for skin, K for kidneys, L for liver, R for the rest of the body, GI for gastrointestinal tract, BR for bladder, F for faeces and, U for urine, I for inhalation, D for dermal application, ORAL for oral administration and IV for intravenous bolus.

5.7.1 Kinetics of BaP

Lungs

$$f_{LU}V_{LU}\frac{\partial}{\partial t}C_{VLU}(t) = Q_C C_V(t) + Q_P C_I(t) - \left(Q_C + \frac{Q_P}{P_B}\right)C_{VLU}(t) - PA_{LU}\left[C_{VLU}(t) - \frac{C_{LU}(t)}{P_{LUA}}\right],$$

(5-3)

$$(1 - f_{LU})V_{LU}\frac{\partial}{\partial t}C_{LU}(t) = PA_{LU}\left[C_{VLU}(t) - \frac{C_{LU}(t)}{P_{LUA}}\right],$$
(5-4)

Adipose tissues

$$V_{AT}\frac{\partial}{\partial t}C_{AT}(t) = Q_{AT}C_A(t) - \frac{Q_{AT}}{P_{ATV}}C_{AT}(t),$$
(5-5)

Skin

$$V_S\frac{\partial}{\partial t}C_S(t) = Q_S C_A(t) - \frac{Q_S}{P_{SV}}C_S(t) + K_P S\left[C_D(t) - \frac{C_S(t)}{P_{DV}}\right],$$
(5-6)

Kidneys

142

$$V_K \frac{\partial}{\partial t} C_K(t) = Q_K C_A(t) - \frac{Q_K}{P_{KV}} C_K(t), \tag{5-7}$$

Liver

$$V_L \frac{\partial}{\partial t} C_L(t) = Q_L C_A(t) - \left(\frac{Q_L}{P_{LV}} + \frac{V_{MAX}}{K_M P_{LV}} + K_B \right) C_L(t) + f_{abs}^{ORAL} K_{ORAL} A_{ORAL}(t), \tag{5-8}$$

Blood

$$f_{AB} V_B \frac{\partial}{\partial t} C_A(t) = Q_C C_{VLU}(t) - Q_C C_A(t), \tag{5-9}$$

$$(1 - f_{AB}) V_B \frac{\partial}{\partial t} C_V(t) = C_{IV}(t) - Q_C C_V(t) + \frac{Q_{AT}}{P_{ATV}} C_{AT}(t) + \frac{Q_S}{P_{SV}} C_S(t) + \frac{Q_K}{P_{KV}} C_K(t) +$$

$$\frac{Q_L}{P_{LV}} C_L(t) + \frac{Q_R}{P_{RV}} C_R(t), \tag{5-10}$$

Rest of the body

$$V_R \frac{\partial}{\partial t} C_R(t) = Q_R C_A(t) - \frac{Q_R}{P_{RV}} C_R(t), \tag{5-11}$$

Gastrointestinal tract

$$\frac{\partial}{\partial t} A_{GI}(t) = K_B C_L(t) - K_F A_{GI}(t), \tag{5-12}$$

Faeces

$$\frac{\partial}{\partial t} A_F(t) = K_F A_{GI}(t). \tag{5-13}$$

5.7.2 Kinetics of 3-OHBaP

Lungs

$$f_{LU} V_{LU} \frac{\partial}{\partial t} C_{vlu}(t) = Q_C C_v(t) - Q_C C_{vlu}(t) - PA_{lu} \left[C_{vlu}(t) - \frac{C_{lu}(t)}{P_{lua}} \right], \tag{5-14}$$

$$(1 - f_{LU}) V_{LU} \frac{\partial}{\partial t} C_{lu}(t) = PA_{lu} \left[C_{vlu}(t) - \frac{C_{lu}(t)}{P_{lua}} \right], \tag{5-15}$$

Adipose tissues

$$f_{AT} V_{AT} \frac{\partial}{\partial t} C_{vat}(t) = Q_{AT} C_a(t) - Q_{AT} C_{vat}(t) - PA_{at} \left[C_{vat}(t) - \frac{C_{at}(t)}{P_{atv}} \right], \tag{5-16}$$

$$(1 - f_{AT}) V_{AT} \frac{\partial}{\partial t} C_{at}(t) = PA_{at} \left[C_{vat}(t) - \frac{C_{at}(t)}{P_{atv}} \right], \tag{5-17}$$

Skin

$$V_S \frac{\partial}{\partial t} C_s(t) = Q_S C_a(t) - \frac{Q_S}{P_{sv}} C_s(t), \tag{5-18}$$

Kidneys

$$f_K V_K \frac{\partial}{\partial t} C_{vk}(t) = Q_K C_a(t) - (Q_K + K_{kbr}) C_{vk}(t) - PA_k \left[C_{vk}(t) - \frac{C_k(t)}{P_{kv}} \right], \tag{5-19}$$

$$(1 - f_K) V_K \frac{\partial}{\partial t} C_k(t) = PA_k \left[C_{vk}(t) - \frac{C_k(t)}{P_{kv}} \right], \tag{5-20}$$

Liver

$$V_L \frac{\partial}{\partial t} C_l(t) = Q_L C_a(t) - \left(\frac{Q_l}{P_{lv}} + \frac{V_{max}}{K_m P_{lv}} + K_b \right) C_l(t) + K_{gil} A_{gi}(t) + f_{3OHBAP} \frac{V_{MAX}}{K_M P_{LV}} C_L(t), \tag{5-21}$$

Blood

$$f_{AB} V_B \frac{\partial}{\partial t} C_a(t) = Q_C C_{vlu}(t) - Q_C C_a(t), \tag{5-22}$$

$$(1 - f_{AB}) V_B \frac{\partial}{\partial t} C_v(t) = -Q_C C_v(t) + Q_{AT} C_{vat}(t) + \frac{Q_S}{P_{sv}} C_s(t) + Q_K C_{vk}(t) + \frac{Q_L}{P_{lv}} C_l(t) +$$

$$\frac{Q_R}{P_{rv}} C_r(t), \tag{5-23}$$

Rest of the body

$$V_R \frac{\partial}{\partial t} C_r(t) = Q_R C_a(t) - \frac{Q_R}{P_{rv}} C_r(t), \tag{5-24}$$

Gastrointestinal tract

$$\frac{\partial}{\partial t} A_{gi}(t) = K_b C_l(t) - \left(K_f + K_{gil} \right) A_{gi}(t), \tag{5-25}$$

Faeces

$$\frac{\partial}{\partial t} A_f(t) = K_f A_{gi}(t), \tag{5-26}$$

Bladder

$$\frac{\partial}{\partial t} A_{br}(t) = K_{kbr} C_{vk}(t) - K_u A_{br}(t), \tag{5-27}$$

Urine

$$\frac{\partial}{\partial t} A_u(t) = K_u A_{br}(t). \tag{5-28}$$

5.8 References

1. IARC (2010) IARC monographs on the evaluation of carcinogenic risks to humans. Ingested nitrate and nitrite, and cyanobacterial peptide toxins. IARC Monogr Eval Carcinog Risks Hum 94:v-vii, 1-412

2. Bouchard M, Gosselin NH, Brunet RC, Samuel O, Dumoulin MJ, Carrier G (2003) A toxicokinetic model of malathion and its metabolites as a tool to assess human exposure and risk through measurements of urinary biomarkers. Toxicol Sci 73 (1):182-194. doi:10.1093/toxsci/kfg061

3. Berthet A, Bouchard M, Valcke M, Heredia-Ortiz R (2012) Using a toxicokinetic modeling approach to determine Biological Reference Values (BRVs) and to assess human exposure to pesticides. . In: Milan J (ed) Impact of Pesticides. Academy Publish,

4. Moir D, Viau A, Chu I, Law FC, Krishnan K (1996) A PBPK model for benzo(a)pyrene in the rat. European Journal of Pharmaceutical Sciences 2 (Supplement 1):92

5. Crowell SR, Amin SG, Anderson KA, Krishnegowda G, Sharma AK, Soelberg JJ, Williams DE, Corley RA (2011) Preliminary physiologically based pharmacokinetic models for benzo[a]pyrene and dibenzo[def,p]chrysene in rodents. Toxicol Appl Pharmacol 257 (3):365-376. doi:Doi 10.1016/J.Taap.2011.09.020

6. Mackay D, Bobra A, Shiu WY, Yalkowsky SH (1980) Relationships between Aqueous Solubility and Octanol-Water Partition-Coefficients. Chemosphere 9 (11):701-711. doi:Doi 10.1016/0045-6535(80)90122-8

7. Mallon BJ, Harrison FL (1984) Octanol-water partition coefficient of benzo(a)pyrene: measurement, calculation, and environmental implications. Bull Environ Contam Toxicol 32 (3):316-323

8. Peters SA (2011) Physiologically based pharmacokinetic (PBPK) modeling and simulations : principles, methods, and applications in the pharmaceutical industry. Wiley, Hoboken, N.J.

9. Marie C, Bouchard M, Heredia-Ortiz R, Viau C, Maitre A (2010) A toxicokinetic study to elucidate 3-hydroxybenzo(a)pyrene atypical urinary excretion profile following intravenous injection of benzo(a)pyrene in rats. J Appl Toxicol 30 (5):402-410. doi:10.1002/jat.1511

10. Heredia-Ortiz R, Bouchard M, Marie-Desvergne C, Viau C, Maitre A (2011) Modeling of the internal kinetics of benzo(a)pyrene and 3-hydroxybenzo(a)pyrene biomarker from rat data. Toxicol Sci 122 (2):275-287. doi:10.1093/toxsci/kfr135

11. Hayes AW (2008) Principles and methods of toxicology. 5th edn. CRC Press, Boca Raton

12. Brown RP, Delp MD, Lindstedt SL, Rhomberg LR, Beliles RP (1997) Physiological parameter values for physiologically based pharmacokinetic models. Toxicol Ind Health 13 (4):407-484

13. Rubinstein RY (1981) Simulation and the Monte Carlo method. Wiley series in probability and mathematical statistics. Wiley, New York

14. Dartois C, Brendel K, Comets E, Laffont CM, Laveille C, Tranchand B, Mentre F, Lemenuel-Diot A, Girard P (2007) Overview of model-building strategies in population PK/PD analyses: 2002-2004 literature survey. British journal of clinical pharmacology 64 (5):603-612. doi:10.1111/j.1365-2125.2007.02975.x

15. Metropolis N, Ulam S (1949) The Monte Carlo Method. J Am Stat Assoc 44 (247):335-341. doi:Doi 10.2307/2280232

16. Manno I (1999) Introduction to the Monte-Carlo method. Akadémiai Kiadó, Budapest

17. Becka M, Urfer W (1996) Statistical aspects of inhalation toxicokinetics. Environ Ecol Stat 3 (1):51-64. doi:Doi 10.1007/Bf00577322

18. Drossel B (2001) Biological evolution and statistical physics. Adv Phys 50 (2):209-295. doi:Doi 10.1080/00018730110041365

19. Bustad A, Terziivanov D, Leary R, Port R, Schumitzky A, Jelliffe R (2006) Parametric and nonparametric population methods - Their comparative performance in analysing a

146

clinical dataset and two Monte Carlo simulation studies. Clin Pharmacokinet 45 (4):365-383. doi:Doi 10.2165/00003088-200645040-00003

20. Csajka C, Verotta D (2006) Pharmacokinetic-pharmacodynamic modelling: history and perspectives. J Pharmacokinet Pharmacodyn 33 (3):227-279. doi:10.1007/s10928-005-9002-0

21. Bauer RJ, Guzy S, Ng C (2007) A survey of population analysis methods and software for complex pharmacokinetic and pharmacodynamic models with examples. The AAPS journal 9 (1):E60-83. doi:10.1208/aapsj0901007

22. Bevington PR, Robinson DK (2003) Data reduction and error analysis for the physical sciences. 3rd edn. McGraw-Hill, Boston

23. Bouchard M, Viau C (1996) Urinary excretion kinetics of pyrene and benzo(a)pyrene metabolites following intravenous administration of the parent compounds or the metabolites. Toxicol Appl Pharmacol 139 (2):301-309. doi:10.1006/taap.1996.0169

24. Cao D, Yoon CH, Shin BS, Kim CH, Park ES, Yoo SD (2005) Effects of aloe, aloesin, or propolis on the pharmacokinetics of benzo[a]pyrene and 3-OH-benzo[a]pyrene in rats. J Toxicol Environ Health A 68 (23-24):2227-2238. doi:10.1080/15287390500182164

25. Weyand EH, Bevan DR (1986) Benzo(a)pyrene disposition and metabolism in rats following intratracheal instillation. Cancer Res 46 (11):5655-5661

26. Ramesh A, Greenwood M, Inyang F, Hood DB (2001) Toxicokinetics of inhaled benzo[a]pyrene: plasma and lung bioavailability. Inhal Toxicol 13 (6):533-555. doi:10.1080/08958370118859

27. Payan JP, Lafontaine M, Simon P, Marquet F, Champmartin-Gendre C, Beydon D, Wathier L, Ferrari E (2009) 3-Hydroxybenzo(a)pyrene as a biomarker of dermal exposure to benzo(a)pyrene. Arch Toxicol 83 (9):873-883. doi:10.1007/s00204-009-0440-0

28. Jongeneelen FJ, Leijdekkers CM, Henderson PT (1984) Urinary excretion of 3-hydroxy-benzo[a]pyrene after percutaneous penetration and oral absorption of benzo[a]pyrene in rats. Cancer letters 25 (2):195-201

29. Moir D, Viau A, Chu I, Withey J, McMullen E (1998) Pharmacokinetics of benzo[a]pyrene in the rat. J Toxicol Environ Health A 53 (7):507-530

30. Ciffroy P, Tanaka T, Johansson E, Brochot C (2011) Linking fate model in freshwater and PBPK model to assess human internal dosimetry of B(a)P associated with drinking water. Environ Geochem Health 33 (4):371-387. doi:10.1007/s10653-011-9382-6

31. Haddad S, Withey J, Lapare S, Law F, Krishnan K (1998) Physiologically-based pharmacokinetic modeling of pyrene in the rat. Environ Toxicol Pharmacol 5 (4):245-255

32. Jongeneelen F, ten Berge W (2012) Simulation of urinary excretion of 1-hydroxypyrene in various scenarios of exposure to polycyclic aromatic hydrocarbons with a generic, cross-chemical predictive PBTK-model. Int Arch Occup Environ Health 85 (6):689-702. doi:10.1007/s00420-011-0713-9

33. Shu HP, Bymun EN (1983) Systemic excretion of benzo(a)pyrene in the control and microsomally induced rat: the influence of plasma lipoproteins and albumin as carrier molecules. Cancer Res 43 (2):485-490

34. Sugihara N, James MO (2003) Binding of 3-hydroxybenzo[a]pyrene to bovine hemoglobin and albumin. J Biochem Mol Toxicol 17 (4):239-247. doi:10.1002/jbt.10084

35. Sagredo C, Mollerup S, Cole KJ, Phillips DH, Uppstad H, Ovrebo S (2009) Biotransformation of benzo[a]pyrene in Ahr knockout mice is dependent on time and route of exposure. Chem Res Toxicol 22 (3):584-591. doi:10.1021/tx8003664

36. Wang JJ, Frazer DG, Stone S, Goldsmith T, Law B, Moseley A, Simpson J, Afshari A, Lewis DM (2003) Urinary benzo[a]pyrene and its metabolites as molecular biomarkers of asphalt fume exposure characterized by microflow LC coupled to hybrid quadrupole time-of-flight mass spectrometry. Anal Chem 75 (21):5953-5960. doi:10.1021/ac030017a

37. Agency for Toxic Substances and Disease Registry (1995) Toxicological profile for polycyclic aromatic hydrocarbons. Rev. edn. U.S. Department of Health and Human Services, Atlanta, GA

148

38. Molliere M, Foth H, Kahl R, Kahl GF (1987) Metabolism of benzo[a]pyrene in the combined rat liver--lung perfusion system. Toxicology 45 (2):143-154

39. Molliere M, Foth H, Kahl R, Kahl GF (1987) Comparison of benzo(a)pyrene metabolism in isolated perfused rat lung and liver. Arch Toxicol 60 (4):270-277

40. DiGiovanni J, Miller DR, Singer JM, Viaje A, Slaga TJ (1982) Benzo(a)pyrene metabolism in primary cultures of mouse epidermal cells and untransformed and transformed epidermal cell lines. Cancer Res 42 (7):2579-2586

41. Chipman JK, Bhave NA, Hirom PC, Millburn P (1982) Metabolism and excretion of benzo[a]pyrene in the rabbit. Xenobiotica 12 (6):397-404

42. van Schooten FJ, Moonen EJ, van der Wal L, Levels P, Kleinjans JC (1997) Determination of polycyclic aromatic hydrocarbons (PAH) and their metabolites in blood, feces, and urine of rats orally exposed to PAH contaminated soils. Arch Environ Contam Toxicol 33 (3):317-322

43. Moore BP, Hicks RM, Knowles MA, Redgrave S (1982) Metabolism and binding of benzo(a)pyrene and 2-acetylaminofluorene by short-term organ cultures of human and rat bladder. Cancer Res 42 (2):642-648

44. Berthet A, Heredia-Ortiz R, Vernez D, Danuser B, Bouchard M (2012) A detailed urinary excretion time course study of captan and folpet biomarkers in workers for the estimation of dose, main route-of-entry and most appropriate sampling and analysis strategies. Ann Occup Hyg 56 (7):815-828. doi:10.1093/annhyg/mes011

45. Foth H, Kahl R, Kahl GF (1988) Pharmacokinetics of low doses of benzo[a]pyrene in the rat. Food and chemical toxicology : an international journal published for the British Industrial Biological Research Association 26 (1):45-51

46. Jongeneelen FJ, Leijdekkers CM, Bos RP, Theuws JL, Henderson PT (1985) Excretion of 3-hydroxy-benzo(a)pyrene and mutagenicity in rat urine after exposure to benzo(a)pyrene. J Appl Toxicol 5 (5):277-282

47. Ramesh A, Inyang F, Hood DB, Archibong AE, Knuckles ME, Nyanda AM (2001) Metabolism, bioavailability, and toxicokinetics of benzo(alpha)pyrene in F-344 rats following oral administration. Exp Toxicol Pathol 53 (4):275-290

48. Wiersma DA, Roth RA (1983) Total body clearance of circulating benzo(a)pyrene in conscious rats: effect of pretreatment with 3-methylcholanthrene and the role of liver and lung. The Journal of pharmacology and experimental therapeutics 226 (3):661-667

49. Jongeneelen FJ, Bos RP, Anzion RB, Theuws JL, Henderson PT (1986) Biological monitoring of polycyclic aromatic hydrocarbons. Metabolites in urine. Scand J Work Environ Health 12 (2):137-143

50. Likhachev AJ, Beniashvili D, Bykov VJ, Dikun PP, Tyndyk ML, Savochkina IV, Yermilov VB, Zabezhinski MA (1992) Biomarkers for individual susceptibility to carcinogenic agents: excretion and carcinogenic risk of benzo[a]pyrene metabolites. Environ Health Perspect 98:211-214

51. Lee W, Shin HS, Hong JE, Pyo H, Kim Y (2003) Studies on the analysis of benzo(a)pyrene and its metabolites in biological samples by using high performance liquid chromatograpy/fluorescence detection and gas chromatography/mass spectrometry. B Kor Chem Soc 24 (5):559-565

52. Hundley SG, Freudenthal RI (1977) A comparison of benzo(a)pyrene metabolism by liver and lung microsomal enzymes from 3-methylcholanthrene-treated rhesus monkeys and rats. Cancer Res 37 (9):3120-3125

53. Prough RA, Patrizi VW, Okita RT, Masters BS, Jakobsson SW (1979) Characteristics of benzo(a)pyrene metabolism by kidney, liver and lung microsomal fractions from rodents and humans. Cancer Res 39 (4):199-206

5.9 Captions to figures

Figure 5-1 : Conceptual PBPK model of the kinetics of BaP and 3-OHBaP in rats. Tissue-diffusion limited processes were considered for BaP and 3-OHBaP in lungs as well as 3-OHBaP in kidneys and adipose tissues

Figure 5-2 : Monte Carlo algorithm proposed for the determination of the parameter values of the tissue: blood partition coefficients, tissue permeability coefficients and metabolism constants in the PBPK model

Figure 5-3 : Comparison of model simulations (lines) with experimental data (symbols) of Marie et al. [9] on the time profiles of BaP in blood and tissues of male Sprague-Dawley rats following an intravenous injection of 40 µmol/kg bw of BaP. Each symbol represents experimental means and vertical bars are standard deviations (n = 4). ● and solid line, lungs; ▲ and dash-dotted line, adipose tissues; ■ and dash-dot-dotted line, blood; ◆ and small-dashed line, liver; ▼ and dotted line, kidney;   and large dashed line, skin

Figure 5-4 : Comparison of model simulations (lines) with experimental data (symbols) of Marie et al. [9] on the time profiles of 3-OHBaP in blood and tissues of male Sprague-Dawley rats following an intravenous injection of 40 µmol/kg bw of BaP. Each symbol represents experimental means and vertical bars are standard deviations (n = 4). ● and solid line, lungs; ▲ and dash-dotted line, adipose tissues; ■ and dash-dot-dotted line, blood; ◆ and small-dashed line, liver; ▼ and dotted line, kidney;   and large dashed line, skin

Figure 5-5 : Comparison of model simulations (lines) with experimental data (symbols) of Bouchard et al. [23] (● and solid line) and Cao et al. [24] (▲ and dashed line) on the time profiles of 3-OHBaP in urine of male Sprague-Dawley rats following an intravenous

151

injection of 40 μmol/kg bw of BaP and 1 mg/kg bw of BaP, respectively. Each symbol represents experimental means and vertical bars are standard deviations (n = 4)

Figure 5-6 : Comparison of model simulations (lines) with experimental data (symbols) of Weyand and Bevan [25] (● and dashed line) and Ramesh et al. [26] (▲ and solid line) on the time profiles of 3-OHBaP in blood following an inhalation of 1 mg/kg bw of BaP in male Sprague-Dawley rats and 100 mg/m³ BaP concentration in male Fisher 344 rats, respectively. Each symbol represents experimental means and vertical bars are standard deviations (n = 3 and 5, respectively)

Figure 5-7 : Comparison of model simulations (lines) with experimental data (symbols) of Payan et al. [27] (right axis; ● and solid line) and Jongeneelen et al. [28] (left axis; ▲ and dashed line) on the time profiles of 3-OHBaP in urine following a dermal application of 70 mg of BaP in male Sprague-Dawley rats and 50 μmol/kg bw of BaP in male Wistar rats, respectively. Each symbol represents experimental means and vertical bars are standard deviations (n = 6 and 3, respectively)

Figure 5-8 : Comparison of model simulations (lines) with experimental data (symbols) of Cao et al. [24] on the time profiles of BaP (● and solid line) and 3-OHBaP (▲ and dashed line) in blood of male Sprague-Dawley rats following an oral exposure of 20 mg/kg bw of BaP. Each symbol represents experimental means and vertical bars are standard deviations (n = 4)

152

5.10 Tables

Table 5-1 : Experimental studies used for the determination of model parameters.

Determined parameters	Compound	Dose	Dosage	Species	Route of exposure	n	Time (h)	Biological matrix	Reference
Metabolic constants, tissue: blood partition coefficients and tissue permeability coefficients	BaP and 3-OHBaP	40 µmol/kg	Single	Male Sprague-Dawley rats	IV	4	0-72	Blood, liver, kidneys, lungs, adipose tissues, skin, urine and faeces	Marie et al [9]
Absorption rates and fractions	BaP	1 µg/kg	Single	Male Sprague-Dawley rats	Intratracheal	3	0-6	Blood	Weyand et Bevan [25]
	BaP	70 µg	Single	Male Sprague-Dawley rats	Cutaneous	6	0-80	Urine	Payan et al. [27]
	BaP	20 mg/kg	Single	Male Sprague-Dawley rats	Oral	4	0-34	Blood	Cao et al. [24]

Table 5-2 : Experimental studies used to assess the model's ability to simulate various animal data available.

Compound	Dose	Dosage	Species	Route of exposure	n	Time (h)	Biological matrix	Reference
BaP and 3-OHBaP	40 μmol/kg	Single	Male Sprague-Dawley rats	IV	4	0 -72	Urine	Bouchard et Viau [23]
BaP	1.7, 3.2, 4 nmol	Single	Male Sprague-Dawley rats	IV et oral	-	0- 35	Blood	Foth et al. [45]
BaP and 3-OHBaP	10, 20, 50 μmol/kg	Three consecutive days	SPF Male Wistar rats	Oral	3	0-144	Urine	Jongeneelen et al. [46]
BaP	100 mg/kg	Single	Fisher-344 rats	Oral	5	0-72	Blood, liver, lungs and faeces	Ramesh et al. [47]
BaP	117 nmol	Single	Male Sprague-Dawley rats	IA et IV	6	0-5	Blood	Weirsma et Roth [48]
BaP and 3-OHBaP	10, 20, 50 μmol/kg	Three consecutive days	Male Wistar rats	Oral and cutaneous	3	0-144	Urine	Jongeneelen et al. [49]
BaP	2, 6, 15 mg/kg	Single	Male Wistar rats	IV	4	0-32	Blood, adipose tissue, kidneys,	Moir et al. [29]

Compound	Dose	Dosage	Species	Route of exposure	n	Time (h)	Biological matrix	Reference
BaP and 3-OHBaP	10 mg/kg	Single	LIO rats	IP	19	0-360	Urine et faeces, liver and lungs	Likhachev et al. [50]
BaP and 3-OHBaP	10 mg/kg	Single	Male Sprague-Dawley rats	IP	5	0-100	Urine	Lee et al. [51]
BaP and 3-OHBaP	1, 20 mg/kg	Single	Male Sprague-Dawley rats	IV et oral	4	0-34	Blood (3-OHBaP), urine et faeces	Cao et al. [24]
3-OHBaP	70 µg	Single	Male Sprague-Dawley rats	Cutaneous	6	0-80	Urine	Payan et al. [27]
BaP	100 µg/m³	Single	Fisher-344 rats	Inhalation	5	0-4	Blood	Ramesh et al. [26]
BaP	1 µg/kg	Single	Male Sprague-Dawley rats	IV	3	0-6	Blood	Weyand et Bevan [25]

Table 5-3 : Model parameters obtained from *in vivo* experimental data of Marie et al. [9], Weyand and Bevan [25], Payan et al. [27] and Cao et al. [24].

			BaP		3-OHBaP	
Matrix		**Parameter**	**Value**	**Sensitivity***	**Value**	**Sensitivity***
Partition coefficients	Lungs	P_{LLA}	2670.00	±15%	2.92	±15%
	Adipose tissues	P_{ATV}	65.90	±15%	1.42	±15%
	Skin	P_{SV}	1.87	±15%	0.80	±15%
	Kidneys	P_{KV}	2.08	±15%	40.40	±15%
	Liver	P_{LV}	12.90	±15%	1.83	±15%
	Rest of the body	P_{RV}	10.00	±15%	1.00	±8.3%
Permeability coefficients	Lungs	PA_{LU} [mL/h]	80.70	±15%	0.20	±15%
	Adipose Tissue	PA_{AT} [mL/h]	-	-	0.711	±15%
	Kidneys	PA_K [mL/h]	-	-	12.90	±15%
Metabolic Constants	Total metabolites	$\sum_{VI} \frac{v_{max}^{(i)}}{K_m^{(i)}}$ [mL/h]	933.00	±3.6%	36.40	±5.8%

Matrix		Parameter	BaP		3-OHBaP	
			Value	Sensitivity*	Value	Sensitivity*
Fraction of 3-OHBaP		$\left(\dfrac{V_{max}^{(3-OHBaP)}}{K_m^{(3-OHBaP)}}\right)\left(\displaystyle\sum_{VI}\dfrac{V_{max}^{(i)}}{K_m^{(i)}}\right)$	0.185	±1.6%	-	-
Elimination rates	Urine	K_B [1/h]	0.338	±15%	42.70	±2.5%
		K_{KB} [1/h]	-	-	0.633	±1.6%
		K_{BU} [1/h]	-	-	0.102	±2.5%
	Faeces	K_F [1/h]	0.334	±15%	0.173	±3%
		K_{GIL} [1/h]	-	-	0.0069	±13%
					3	
Absorption constants	Oral	$f_{abs}K_{ORA}$ [1/h]	2330.00	±15%	-	-
	Cutaneous	k_P [cm/h]	0.0119	±15%	-	-
		P_{DV}	1.0	±15%	-	-
	Inhalation	P_B	2.04	±5.5%	-	-

* Range of parameter variation during Monte Carlo simulation to obtain 91.7 ± 1.63% of runs (n~O(10^3)) within a maximum variation of ±10% in the simulated urinary excretion profiles compared to default parameter values.

5.11 Figures

Figure 5-1: Conceptual PBPK model of the kinetics of BaP and 3-OHBaP in rats.

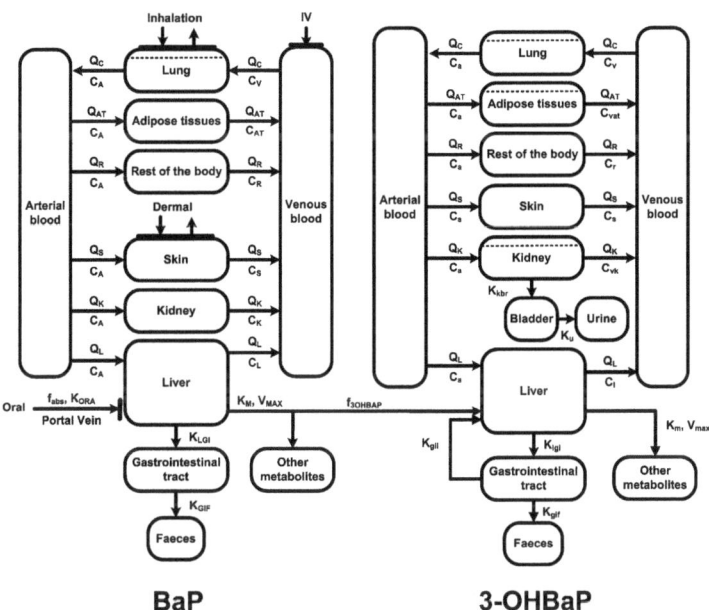

Figure 5-2: Monte Carlo algorithm proposed for the determination of the parameter values of the tissue: blood partition coefficients, tissue permeability coefficients and metabolism constants in the PBPK model

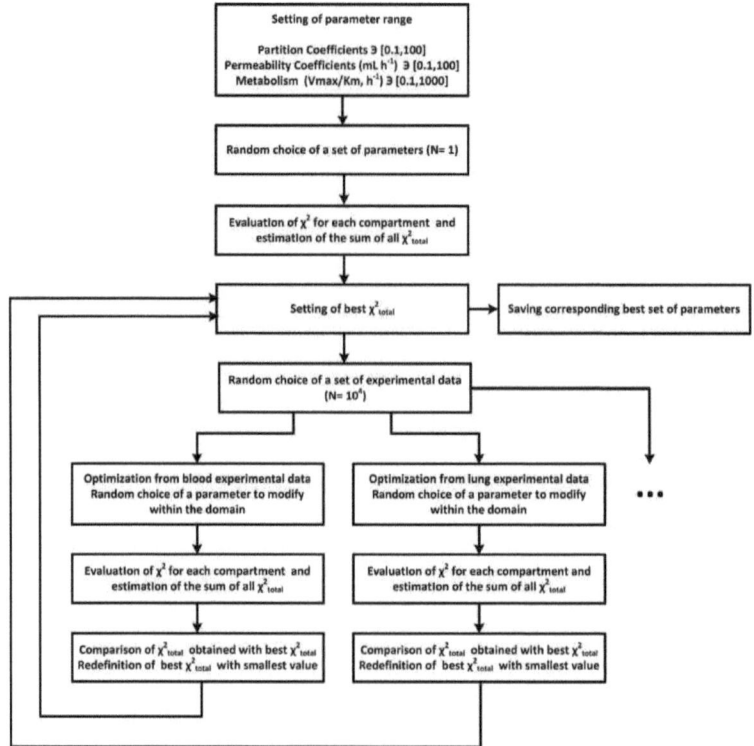

Figure 5-3: Comparison of model simulations (lines) with experimental data (symbols) of Marie et al. [9] on the time profiles of BaP in blood and tissues of male Sprague-Dawley rats following an intravenous injection of 40 µmol/kg bw of BaP.

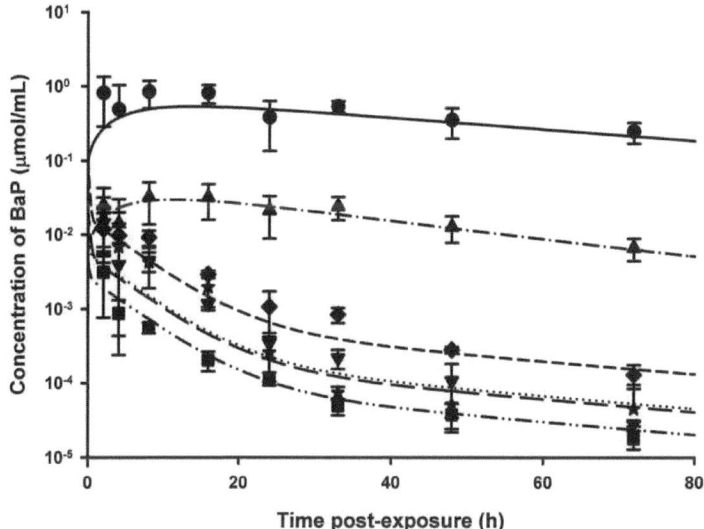

Figure 5-4: Comparison of model simulations (lines) with experimental data (symbols) of Marie et al. [9] on the time profiles of 3-OHBaP in blood and tissues of male Sprague-Dawley rats following an intravenous injection of 40 μmol/kg bw of BaP.

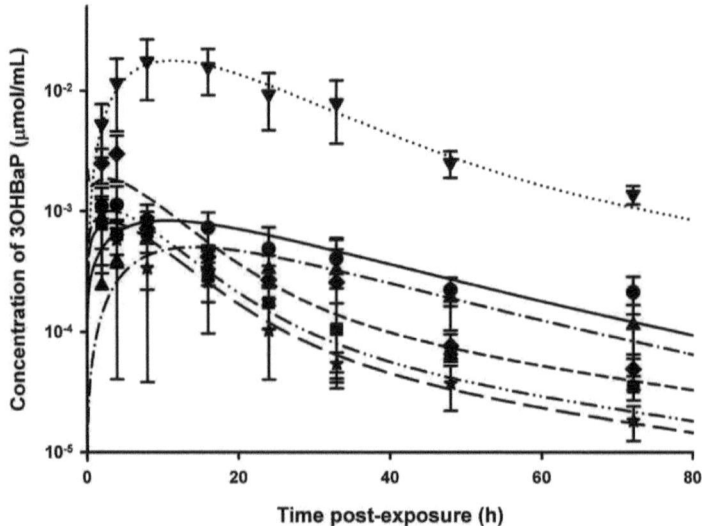

Figure 5-5: Comparison of model simulations (lines) with experimental data (symbols) of Bouchard et al. [23] (● and solid line) and Cao et al. [24] (▲ and dashed line) on the time profiles of 3-OHBaP in urine of male Sprague-Dawley rats following an intravenous injection of 40 μmol/kg bw of BaP and 1 mg/kg bw of BaP, respectively.

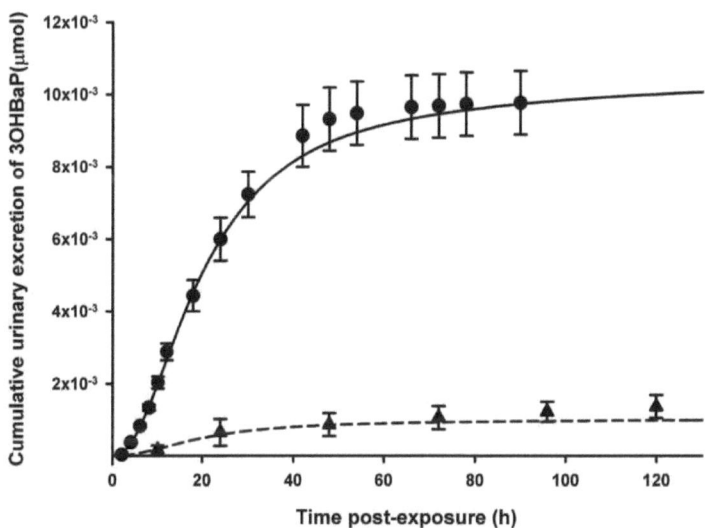

Figure 5-6: Comparison of model simulations (lines) with experimental data (symbols) of
Weyand and Bevan [25] (● and dashed line) and Ramesh et al. [26] (▲ and solid line) on
the time profiles of 3-OHBaP in blood following an intratracheal instillation of 1 μg/kg
bw of BaP in male Sprague-Dawley rats and 100 mg/m³ BaP concentration in male Fisher
344 rats, respectively.

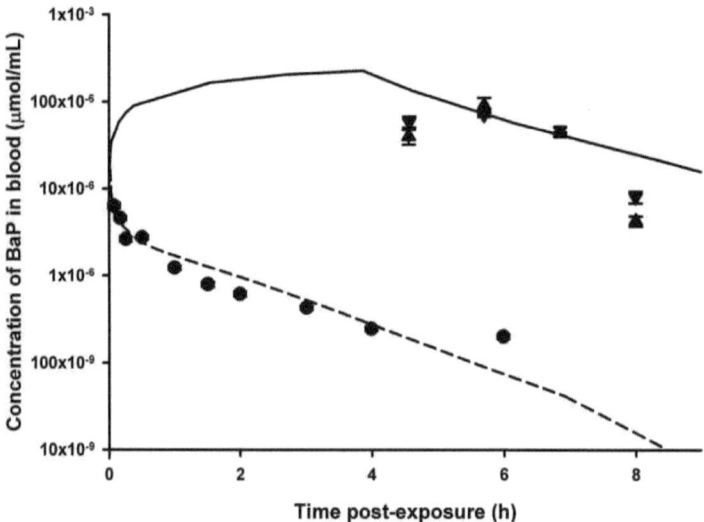

Figure 5-7: Comparison of model simulations (lines) with experimental data (symbols) of Payan et al. [27] (right axis; ● and solid line) and Jongeneelen et al. [28] (left axis; ▲ and dashed line) on the time profiles of 3-OHBaP in urine following a dermal application of 70 µg of BaP in male Sprague-Dawley rats and 50 µmol/kg bw of BaP in male Wistar rats, respectively.

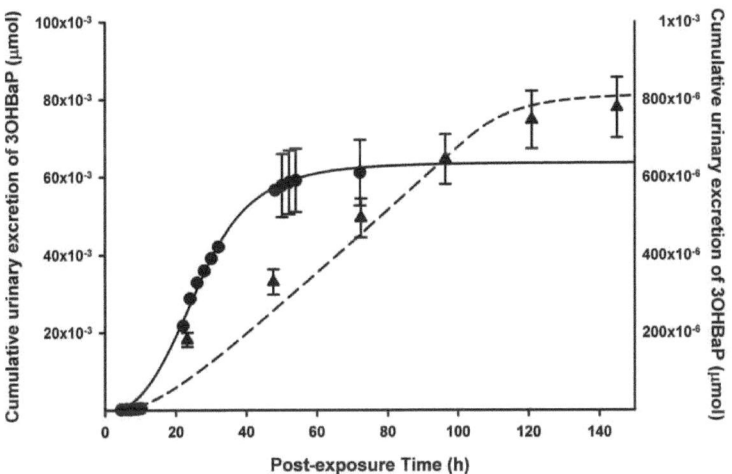

Figure 5-8: Comparison of model simulations (lines) with experimental data (symbols) of Cao et al. [24] on the time profiles of BaP (● and solid line) and 3-OHBaP (▲ and dashed line) in blood of male Sprague-Dawley rats following an oral exposure of 20 mg/kg bw of BaP.

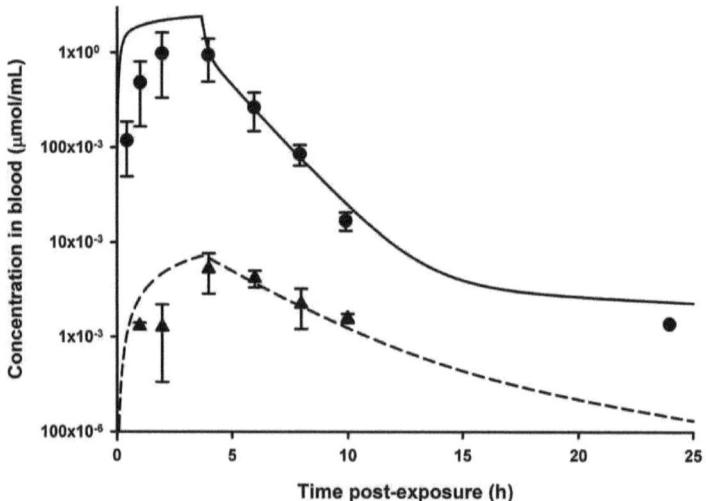

6 Troixème article : Use of physiologically-based pharmacokinetic and simple compartmental toxicokinetic modeling to simulate the time course of 3-hydroxybenzo(a)pyrene metabolite in workers exposed to polycyclic aromatic hydrocarbons

Roberto Heredia-Ortiz, Anne Maître, Damien Barbeau, Michel Lafontaine, Michèle Bouchard.

Manuscript accepted for publication in PLOS ONE on May 20th, 2014.

Use of physiologically-based pharmacokinetic modeling to simulate the profiles of 3-hydroxybenzo(a)pyrene in workers exposed to polycyclic aromatic hydrocarbons

Roberto Heredia-Ortiz[1], Anne Maître[2], Damien Barbeau[2], Michel Lafontaine, and Michèle Bouchard[1,*]

[1] Département de santé environnementale et santé au travail, Chaire d'analyse et de gestion des risques toxicologiques and Institut de recherche en santé publique de l'Université de Montréal (IRSPUM), Université de Montréal, C.P. 6128, Succursale Centre-ville, Montréal, Québec, Canada, H3C 3J7

[2] EPSP, Equipe environnement et prédiction de la santé des populations, Laboratoire TIMC (UMR 5525), CHU de Grenoble, Université Joseph Fourier, Domaine de la Merci, 38700 La Tronche, France.

* Corresponding author:
Michèle Bouchard
Associate professor
Department of Environmental and Occupational Health
University of Montreal
P.O Box 6128, Main Station
Montreal (QC),
H3C 3J7, CANADA
Telephone number: (514) 343-6111 ext 1640
Fax number: (514) 343-2200

6.1 Abstract

Biomathematical modeling has become an important tool to assess xenobiotic exposure in humans. In the present study, we have used a human physiologically-based pharmacokinetic (PBPK) model and an simple compartmental toxicokinetic model of benzo(a)pyrene (BaP) kinetics and its 3-hydroxybenzo(a)pyrene (3-OHBaP) metabolite to reproduce the time-course of this biomarker of exposure in the urine of industrially exposed workers and in turn predict the most plausible exposure scenarios. The models were constructed from in vivo experimental data in rats and then extrapolated from animals to humans after assessing and adjusting the most sensitive model parameters as well as species specific physiological parameters. Repeated urinary voids from workers exposed to polycyclic aromatic hydrocarbons (PAHs) have been collected over the course of a typical workweek and during subsequent days off work; urinary concentrations of 3-OHBaP were then determined. Based on the information obtained for each worker (BaP air concentration, daily shift hours, tasks, protective equipment), the time courses of 3-OHBaP in the urine of the different workers have been simulated using the PBPK and toxicokinetic models, considering the various possible exposure routes, oral, dermal and inhalation. Both models were equally able to closely reproduce the observed time course of 3-OHBaP in the urine of workers and predicted similar exposure scenarios. Simulations of various scenarios suggest that the workers under study were exposed mainly by the dermal route. Comparison of measured air concentration levels of BaP with simulated values needed to obtain a good approximation of observed time course further pointed out that inhalation was not the main route of exposure for most of the studied workers. Both kinetic models appear as a useful tool to interpret biomonitoring data of PAH exposure on the basis of 3-OHBaP levels.

6.2 Introduction

Polycyclic aromatic hydrocarbons (PAHs) are a class of ubiquitous contaminants found in many industrial settings such as aluminum plants, silicon production plants and creosote impregnation plants [1,2]. Workers are exposed through inhalation but also dermal contact depending on their job category, task and protective equipment [3-5]. Several members of this class of compounds have been categorized as probable or possible human carcinogens and the most studied PAH, benzo(a)pyrene (BaP), has been classified as a human carcinogen by the International Agency for Research on Cancer (IARC) [6].

Most industrial settings perform a strict monitoring of air levels of PAHs in their facilities, as well as skin patch analysis and/or in many cases rely on the biomonitoring of exposure of their staff [7,42-46]. Some years ago, Jongeneelen et al. [8,9] proposed the use of 1-hydroxypyrene, a metabolite of the non-carcinogen PAH pyrene, as a biomarker of exposure to PAHs. More recently, 3-hydroxybenzo(a)pyrene (3-OHBaP) has been proposed as a complementary measure to better assess carcinogenic PAH exposure [7,10-14]. Typical urinary levels of 3-OHBaP in workers have been reported to be around 0.5 nmol/mol creatinine while the general population values vary around 0.1 nmol/mol creatinine [10]. Campo et al. [15] assessed PAH exposure in coke-oven workers by determining urinary profiles of hydroxylated (including 3-OHBaP) and unmetabolized PAHs showing that both hydroxylated metabolites and unmetabolized PAHs in urine were useful biomarkers of exposure to PAHs. A few years ago, Forster et al. [10] assessed external and internal exposure to PAHs in 255 workers of different industries and the reliability of 3-OHBaP as a biomarker of internal exposure; BaP and 3-OHBaP were detected in the urine of workers from all workplaces. Also, positive correlations (r = 0.6 to 0.9) between urinary 3-OHBaP and 1-OHP as well as the sum of hydroxyphenols were found in workers of coke and graphite electrode production plants.

172

To help interpretation of biomonitoring data, toxicokinetic models have been proposed. Biologically-based toxicokinetic models allowing to relate the time course of a biomarker of exposure to the time-varying amounts in the body and doses per unit of time have been used by our group to reconstruct exposure in workers exposed to methanol and pesticides [16-19]. Recently, such type of toxicokinetic model has been developed to describe the kinetics of BaP and its 3-OHBaP biomarker of exposure in animals and humans [20]. Other types of kinetic models include physiologically-based pharmacokinetic (PBPK) models. Typically, these types of models are constructed from the interactions observed for a given substance with different organs, tissues and fluids *in vitro* [21-23], directly from the physicochemical characteristics of the substance under study [24] or simply by mathematically extrapolating its behaviour on the basis of similar compounds [25,26]. An animal PBPK model of BaP and its 3-OHBaP metabolite was recently developed to describe the kinetics of BaP and 3-OHBaP [27]. The objective of our study was to extrapolate this PBPK model to humans and verify its use to reproduce time courses of 3-OHBaP in the urine of workers and predict the most plausible exposure scenarios. PBPK model predictions were also compared with those of a simple human toxicokinetic model based on rat data and relating urinary excretion of 3-OHBaP to whole body levels and BaP doses by different routes of exposure.

173

6.3 Materials and Methods

6.3.1 Establishment of a human PBPK model extrapolated from a rat PBPK model

6.3.1.1 Conceptual and functional representation

A rat PBPK model representing the kinetics of BaP and its metabolite 3-OHBaP [27] was used as a basis for the establishment of a human-extrapolated model. The conceptual model is shown in Figure 6-1 and its mathematical representation is described in the Appendix. The kinetics of BaP and 3-OHBaP were simulated for three different routes of exposure: respiratory, dermal and oral exposures.

6.3.1.2 Human extrapolated model parameters

The rat physiological parameter values, such as organ weights, blood flow and total ventilation rates, were replaced by human values described in the medical literature by Davies and Morris [28] and Brown et al. [29]. Tissue: blood partition coefficients (describing perfusion-limited transfers) and tissue permeability coefficients (for diffusion-limited processes) were assumed to be species invariant; they were kept as established in the rat PBPK model, hence from in vivo time course data using a set of Monte Carlo algorithms with the Pearson χ^2 as the best fit criteria [27, 30].

The other parameter values, which are known to be species specific, were best-fitted from available human data and include the metabolism rate, the excretion parameters and the dermal absorption parameters. The excretion parameters were adjusted from known rates (biliary excretion, renal filtration) in rats compared to humans [29] as described in Table 6-1. The metabolism rates of BaP and 3-OHBaP have been scaled from the rat model values as described in Table 6-1, and the BaP/3-OHBaP metabolism ratio was kept as obtained in rats. The rat-to-human scaling constant $C_{rat-human}$ is defined as the ratio between the human metabolism rates and the rat metabolism rates. This constant simply reflects the linear scaling that the rat metabolism rates must undertake in order to represent the human metabolism rates. The rat-to-human scaling constant $C_{rat-human}$ (defined as the ratio between human and rat metabolism rates) was computed directly from a least-square best-fit to an observed elimination time course of 3-OHBaP in the urine of a BaP exposed worker (using again the χ^2 statistic as goodness of fit criteria). More specifically, the metabolism rate values of BaP and 3-OHBaP (taken to occur in the liver essentially) were determined by best-fit adjustments to the observed elimination time course of 3-OHBaP in urine over the 35 to 45-h period following the onset of a weekly exposure in worker 4 presented in Figure 6-3, after setting human physiological parameter values and excretion rate values. The dermal absorption parameters were extrapolated as described in Table 6-1 on the basis of Morimoto's scaling [31].

6.3.1.3 Model simulation

Once all the parameters of the model have been fixed, simulations were carried out by numerically solving the system of differential equations representing the kinetics of BaP, on the one hand, and 3-OHBaP, on the other hand. All simulations were performed using Matlab 2010a (Mathworks, MA, USA).

175

6.3.1.4 Sensitivity simulation

Sensitivity analyses were performed to verify the stability of the overall model, by stochastically varying simultaneously all the parameter values of the model [32,33]. This allowed testing the model for different routes of exposure and synergistic, combined and opposing effects of multiple parameter variations. We estimated the difference between the urinary excretion profiles resulting from all the parameter variations (1000 runs) and those obtained with the default parameter values (initially optimized), and calculated the mean percentages (± SD) of 20 data points (t = 8, 16, 24, 32, 40, 48, 56, 64, 72, 80, 88, 96, 104, 112, 120, 128, 136, 144, 152, and 160 h) falling within a 10% variation compared to initially optimized default parameter values. This allowed identifying the most sensitive parameters, as very sensitive parameters result in large variations in the urinary excretion profiles compared to less sensitive parameters, which cause very little urinary changes.

6.3.2 Development of a simple compartmental toxicokinetic model

A simple one-compartment model representing BaP and 3-OHBaP in the whole body was built to relate urinary excretion time courses of 3-OHBaP with BaP absorbed doses by different routes-of-entry (Figure 6-2). The mathematical representation is described in the Appendix. The only key parameters in the model were the absorption rate (k_a) of BaP by the different routes-of-entry (intravenous, $k_{a_{iv}}$, inhalation, $k_{a_{inh}}$; dermal, $k_{a_{der}}$; and oral, $k_{a_{oral}}$), the elimination rate from the body (k_b) and the fraction of dose (α) recovered in urine as 3-OHBaP.

They were established on the basis of the same rat time course data [13,34-36] used to build the rat PBPK model [27]. The Matlab 2010a built-in fit functions were used to determine rat parameter values, all at once, using the analytical solution of the differential equations. Key

176

parameters in this model were the absorption rate by the different routes-of-exposure and the elimination rate from the body of BaP as 3-OHBaP, k_b. Extrapolation of parameter values from rats to humans was heuristically determined as described in Table 6-1. As for the PBPK model, the dermal absorption parameter was extrapolated on the basis of Morimoto's scaling [31] (Table 6-1). Extrapolation of the elimination rate from the body of BaP as 3-OHBaP, k_b, accounted for species-specific changes in the metabolism rates and urinary elimination rate. This elimination rate is also influenced by the clearance of each organ. However, only the biological process contributing the most to the observed time-courses needed a species-specific adjustment, thus in our case the amounts of BaP in adipose tissues (according to Heredia-Ortiz et al. [20]) (see Table 6-1). Given the small number of parameters, no sensitivity analysis was carried out for the single compartment model. Once the parameters of the model were fixed, simulations of the same time courses as those reproduced with the human PBPK model were carried out in Matlab 2010a, using the exact solutions of the differential equations.

6.3.3 PBPK and toxicokinetic model evaluation of 3-OHBaP time course data in workers

We assessed the capacity of our models to simulate urinary excretion time courses of 3-OHBaP in three groups of workers other than the one used to determine the human metabolic constants. Data provided for the time courses of 3-OHBaP in the urine of workers assessed by Lafontaine et al. and Gendre et al. [3,5,37] were simulated with the models. These included data from five subjects exposed in an artificial shooting target factory (including two Occupational Health practitioners on site for the biomonitoring and three full-time workers) and from five workers of a carbon disk brake production plant, respectively. A workshift personal air sampling of vapor and particulate PAHs was performed over two workdays for the assessment of atmospheric BaP concentrations, as described in Lafontaine et al. and Gendre et al. [3,37]. Concentrations of 3-OHBaP were also determined as described in Simon et al. [38] in complete urine voids collected over the course of about a two-day work exposure and subsequent day with limited exposure. In the artificial shooting target factory, none of the workers wore respiratory protection but two of the workers wore gloves regularly. In the carbon disk brake production plant, one of the workers wore a cartridge mask and one wore a paper mask; the other workers did not wear respiratory protection equipment.

The third group monitored by the team of Professor Maître consisted of four employees repairing metallurgical furnaces in a silicon production plant. All workers assessed by the University Joseph-Fourier of Grenoble Hospital, France, gave their informed consent to participate in this study. In France, Occupational Health Physicians are mandated for the follow-up of workers, and perform routine biomonitoring of exposure. All samples were collected under the responsibility of the Occupational Health Physician of the company as part of the routine follow-up of workers and sent to the Grenoble Hospital. For laboratory analysis at the Hospital, personal identification coding were used and treated under the responsibility of a physician and a biologist sworn to medical secrecy. Results for each individual were then sent only to the Occupational Health Physician of the company, who interpreted the results to the workers. Only anonymous data was handed over to the researchers for this study.

Workers were aged between 28 and 54 years old and two of the four workers were smokers; they were assigned to the removal of plates and crowns surrounding the electrodes inside two ovens. This activity causes a significant release of PAHs in the air because electrodes are loaded with coal tar pitch. In addition, operators were in direct and indirect skin contact with the tar pitch. During this activity, no collective protection equipment was in place, but all the workers wore respiratory masks (with type ABEK2P2 or A2P3 cartridges) and leather handling gloves.

The disassembly of plates and crowns lasted four days (Tuesday to Friday), and employees were on a weekly rest during a two-day period before and after this task. A personal air sampling of vapor and particulate PAHs (n =19) was performed on the first and last day of the work week (Tuesday and Friday), in accordance to the French NF X 43-294 standard. Beginning and end-of-shift urine samples were also collected over the course of the work week. During the following 48-h period off work, workers were asked to collect all their micturitions in separate bottles. Polypropylene bottles were used for urine collection and samples were stored at -20 °C until analysis. Concentrations of 3-OHBaP in urine were measured according to the method published by Barbeau et al. [39] and values were corrected for creatinine concentrations.

Model simulation of the observed urinary time courses of 3-OHBaP considering an inhalation and dermal exposure allowed assessing the influence of the route-of-exposure on the time profiles. Inhalation exposure scenarios, as assessed from measured BaP air concentrations and time-of-shifts during a workweek, served as an initial simulation of the time courses of 3-OHBaP in the urine of workers. Ventilation rate was taken to be 7.98 l/min for workers [28,29]. Dermal exposure scenarios were simulated by considering a whole-body dermal dose of the same amounts of BaP diluted in about one liter. Changing BaP concentration or exposed surface only affects proportionally the simulated time-dependent concentration values, without affecting absorption or elimination slopes. Direct oral exposure was considered insignificant in those workers.

As a second independent step, in order to find the exposure doses that best described the urinary excretion profiles, a visual adjustment to the data points was first carried out. Then, a set of approximately a thousand doses around that value was established. A Monte Carlo simulation randomly picked values from this data set and checked for the corresponding $\chi 2$ statistic; at the end of a thousand iterations, the best value for each dose was kept.

6.4 Results

6.4.1 PBPK model extrapolated to humans and sensitivity analysis

The parameter values of the PBPK model are presented in Table 6-2. Using the model structure established from rat data (including entero-hepatic recirculation), the tissue-blood partition coefficients determined from the *in vivo* time-course data in rats, but a metabolism and elimination rate faster than in rats and slower dermal absorption rate, the PBPK model was able to simulate various sets of human time-course data.

Monte Carlo simulations showed that the parameters influencing the most the overall excretion kinetics of 3-OHBaP were the metabolism rate and the elimination rate. Even a very small variation in values of the metabolic constants of BaP and 3-OHBaP can cause the simulation of the expected urinary excretion to change considerably (Table 6-2). In addition to the importance of the metabolic constants, we found that the elimination rates of 3-OHBaP through the kidneys and bladder largely determine the urinary excretion profiles. The fastest 3-OHBaP transfer rate is the transfer from the liver to the bile (K_b more than one orders of magnitude higher than the urinary elimination rate K_{kb}), which leads directly to high levels of 3-OHBaP in faeces. The values of the other excretion parameters are similar, the K_{bu} value for the 3-OHBaP transferred from the bladder to urine being of the same order of magnitude as the fecal excretion K_F for BaP and K_f 3-OHBaP. These simulations also showed that the permeability coefficients are not the key parameters determining the overall time-course curves of BaP or 3-OHBaP in the various tissues. For instance, smaller permeability coefficients would primarily decrease concentrations at all times proportionally in organs limited by diffusion without significantly altering the curves of the other compartments.

6.4.2 Single compartment toxicokinetic model

The toxicokinetic model parameter values for BaP absorption rate (ka) by inhalation (ka_{inh}) and dermal contact (ka_{der}) were respectively estimated to be: 48.35×10^{-3} h^{-1} and 61.81×10^{-3} h^{-1}. The elimination rate from the body (k_b) was equal to 198.26×10^{-3} h^{-1}, while the fractions of dose (α) recovered in urine as 3-OHBaP following inhalation and dermal exposure were found to be 3.99×10^{-2} and 1.64×10^{-2}, respectively.

6.4.3 PBPK and toxicokinetic model evaluation of 3-OHBaP time course data in workers

With the parameter values presented in Table 6-2 and the simulated mostly dermal exposure scenarios (concentrations and time-of-shift) presented in Tables 6-3 and 6-4, the PBPK model was able to reproduce closely the time courses of 3-OHBaP in the urine of individuals exposed to PAHs in two different plants, with complete time-voids over the course of a workweek (Figures 6-3 and 6-4). The model was also able to reproduce the times profiles observed in a third industry where workers provided spot samples prior, during and following a work period (Figure 6-5 and Table 6-5). PBPK model simulations were compared to predictions obtained with a simple compartmental toxicokinetic model and Figures 6-3, 6-4, and 6-5 show that both models provided similar fits to the observed data. For most workers, a more pronounced difference was observed between simulated and observed time courses when considering an exposure by inhalation. Therefore, we have optimized for dermal exposure scenarios, which considered for most workers an exposure not only during their work-shift hours but also a continued absorption after work task hours. Simulations thus suggest a dermal contamination of BaP during work, without a proper cleaning thereafter.

Furthermore, the inhaled dose scenarios simulated to reproduce the urinary profiles were in most cases higher than those recorded from airborne measurements of BaP concentrations in the facilities and time-of-shifts (Tables 6-3, 6-4, and 6-5). Cases in which the simulated inhaled BaP concentration levels needed to obtain a better fits to observed profiles were much higher than measured air concentration values further indicate that inhalation was not the main route of exposure. Therefore, it suggests a mostly dermal exposure for workers of the artificial shooting target factory (workers 1 to 5) and the carbon disk brake production plant (workers 6 to 10, except maybe worker 8 on day one where simulating an inhalation exposure resulted in good predictions) as well as worker 12 of the silicon production industry, evidently, but also possibly some of the other workers (workers 11 and 14). In addition, for workers of the silicone production plant performing oven repairs (Figure 6-5 and Table 6-5), we have considered in the initial modeling that BaP air concentrations on the first three exposure days (Tuesday to Thursday) were the same as those measured on the first exposure day of the week (Tuesday, where atmospheric measurements were available). However, there might be significant variations in BaP concentrations throughout the week as values measured on day four of exposure (Friday) were very different from those measured on the first day of the week (two orders of magnitude).

Interestingly, Figure 6-3 allows comparing 3-OHBaP excretion profiles in Occupational Health practitioners on site to perform the biomonitoring of workers (physician as subject 1 and nurse as subject 2) with those of industry workers. In line with observed time course data, low exposures were simulated at the onset of collection period with a progressive rise above typical general population values by the end of the first exposure day and values reaching those observed in workers by the end of the second day of exposure. On the other hand, the observed time profiles in workers (subjects 3-5) show a background exposure at the onset of workweek (with corresponding urinary values around 0.5 nmol/mol creat.) with a progressive rise over the course of workweek. In workers of the second industry, a similar pattern was observed (Figure 6-4).

182

Figure 6-5 also shows a good adequacy between model simulations and observed data in workers of the third industry. Two of the workers (11 and 12) exhibited similar profiles, with modeled high exposure at the beginning of the workweek and ensuing decreasing exposures throughout the week (with daily peaks and troughs). The most highly exposed worker was also modeled with a clear exposure mainly on day 1 and progressive decrease thereafter (worker 14) whereas worker 13 showing the lowest exposure to BaP exhibited the most erratic profiles (and hence excretion values < 0.4 nmol/mol creat.).

6.5 Discussion

Current risk assessment studies attempt to establish reliable tools enabling to reconstruct absorbed doses or exposure doses in individuals from measurements of exposure biomarkers in accessible biological matrices. In our study, we have ascertained a one-to-one relation between the studied exposure dose and 3-OHBaP biomarker in urine using a thorough knowledge of the toxicokinetics of BaP and 3-OHBaP encoded in two mathematical models, which describe the major determinants of the excretion kinetics.

The development of two types of toxicokinetic models in humans, based on in vivo time course data of BaP and 3-OHBaP in rats, as obtained from the available literature, proved useful to reproduce adequately the excretion time courses of 3-OHBaP biomarker in workers and predict the most plausible exposure scenarios. Although the mathematical representations of the two types of models differ, parameter values of both models were easily extrapolated from animals to humans using extrapolation factors and in turn predicted similar exposures. However, the PBPK model, with a more physiological description of the internal kinetics and based on a rat model evaluated with detailed time course data on the internal kinetics in animals [27], allowed pointing out the critical determinants of the overall excretion kinetics. The rat PBPK model indicated that the metabolic rates of both BaP and 3-OHBaP (with a major contribution of the metabolism from the liver over the lungs and negligible relative skin metabolism) and excretion rate of 3-OHBaP were the most sensitive parameters governing the observed 3-OHBaP urinary excretion kinetics. The rat PBPK model also showed that the skin permeability coefficient and tissue-permeability coefficients for diffusion-limited processes were not the key parameters determining the overall time course curves of BaP or 3-OHBaP in the various tissues; a smaller skin permeability coefficient, as expected in humans compared to rats, would only decrease concentrations at all times proportionally in organs without significantly altering the shape of the curves. The PBPK modeling of rat data [27] further showed that the slow release of BaP from adipose tissues and lungs (possibly lipid components of the lung) was the rate-limiting step driving the overall observed time profiles of BaP in blood, even following a dermal exposure where absorption rate is expected to be slower than following inhalation exposure. By setting these key parameter values of the PBPK model to human specific values, the rat-based PBPK model provided very good fits to the available worker time course data.

Both the PBPK and toxicokinetic modeling of BaP and 3-OHBaP have also shown that inference of the main route-of-entry on the basis of model fitting of observed excretion time courses of 3-OHBaP, considering an inhalation or dermal exposure, is not straightforward. The available number of sampling points to describe the excretion profiles of 3-OHBaP could be reproduced by several plausible exposure scenarios. Therefore, unless one has information on worker tasks or exposure concentrations, the resulting urinary biomarker profiles alone do not allow confirming the main route of BaP exposure. Our model has also shown that, a few hours after exposure, the urinary excretion profiles are governed by the slow release of BaP from organs retaining the parent compound, namely adipose tissues and lungs. A similar observation can be drawn from a toxicokinetic model previously developed by our team [20]. Also, according to this PBPK modeling and prior toxicokinetic modeling [20] based on observed time-course data in rats [35,40], the urinary excretion of 3-OHBaP

184

is delayed with respect to blood profiles. Our modeling assumed that the major reason for such a delay was the relatively slow transfer of 3-OHBaP occurring in the kidneys and a possible interaction of 3-OHBaP in the bladder.

Overall, our kinetic modeling provided a tool to better interpret 3-OHBaP biomonitoring data in workers exposed to PAHs. It pointed out that there is a need for sufficient time course points in workers over a workweek (and ideally, complete urine voids) to be able to reconstruct exposure. In the modeling process, it remains important to consider information on tasks performed by workers and air concentrations, as was done by other authors [3,5,10,37], to better assess potential major route-of-exposure. The use of multiple biomarker measurements, such a 1-hydroxypyrene (1-OHP) and 3-OHBaP in combination, may further help interpreting biomonitoring results, especially given that BaP is a substance that is more representative of carcinogenic PAHs than pyrene, as highlighted by others [10]. In addition, there are differences in the excretion kinetics of 1-OHP and 3-OHBaP [3,40], such that 1-OHP excretion profile is more obviously influenced by the main route-of-entry (inhalation versus dermal) compared with 3-OHBaP. This has been observed in a recent study, in which the urinary excretions of 1-OHP in humans exposed to PAHs have been compared to simulations obtained with a PBPK model for pyrene [41]. In the latter study, simulated urinary profiles presented different time-to-peak levels for each route of exposure: less than 8 h for inhalation in electrode paste workers and around 15 h for volunteers dermally exposed. The PBPK model developed for pyrene [41] differs from current BaP and 3-OHBaP model in the sense that essential parameters such as the tissue-blood partition coefficients were established from in vivo time courses rather than in vitro; specific determinants of BaP and 3-OHBaP kinetics were also accounted for, such as chemical-specific metabolism and atypical renal excretion, and differences in elimination rates and entero-hepatic recirculation.

185

6.6 Acknowledgments

This work was supported by the Agence nationale de sécurité sanitaire de l'alimentation, de l'environnement et du travail (ANSES) [Grant number 2011-CRD-13] and the Chair in Toxicological Risk Assessment and Management of the University of Montreal.

6.7 Appendix

Differential equations defining both models are described. First, the mathematical representation of the PBPK model is presented. As usual, C stands for concentrations (μmol/mL), A for amounts (μmol), Q for regional blood flow rate, V for volumes, P for tissue:blood partition coefficients, PA for permeability-area coefficients, f for fractions (e.g. fraction of blood in tissues), k_P for skin permeability coefficient, K for rates, S surface of exposure, V_{MAX} total maximum rate of metabolism and K_M Michaelis constant. Tissues and fluids are labeled in indices as follows for BaP: LU for lungs, AT for adipose tissues, V for venous blood, A for arterial blood, S for skin, K for kidneys, L for liver, R for the rest of the body, GI for gastrointestinal tract, F for faeces and U for urine. Tissues and fluids are labeled as follows for 3-OHBaP: lu for lungs, at for adipose tissues, v for venous blood, a for arterial blood, s for skin, k for kidneys, l for liver, r for the rest of the body, gi for gastrointestinal tract, br for bladder, f for faeces and u for urine. Second, the mathematical representation of the toxicokinetic model is presented. D is the dose absorbed, k_a are the rates of absorption, k_b is the rate of elimination and α is the fraction of 3-OHBaP eliminated through urine.

6.7.1 PBPK modeling of BaP kinetics

Lungs

$$f_{LU}V_{LU}\frac{\partial}{\partial t}C_{VLU}(t) = Q_CC_V(t) + Q_PC_I(t) - \left(Q_C + \frac{Q_P}{P_B}\right)C_{VLU}(t) - PA_{LU}\left[C_{VLU}(t) - \frac{C_{LU}(t)}{P_{LUA}}\right],$$

$$(6\text{-}1)$$

$$(1 - f_{LU})V_{LU}\frac{\partial}{\partial t}C_{LU}(t) = PA_{LU}\left[C_{VLU}(t) - \frac{C_{LU}(t)}{P_{LUA}}\right], \tag{6-2}$$

Adipose tissues

$$V_{AT}\frac{\partial}{\partial t}C_{AT}(t) = Q_{AT}C_A(t) - \frac{Q_{AT}}{P_{ATV}}C_{AT}(t), \tag{6-3}$$

187

Skin

$$V_S \frac{\partial}{\partial t} C_S(t) = Q_S C_A(t) - \frac{Q_S}{P_{SV}} C_S(t) + k_P S \left[C_D(t) - \frac{C_S(t)}{P_{DV}} \right], \tag{6-4}$$

Kidneys

$$V_K \frac{\partial}{\partial t} C_K(t) = Q_K C_A(t) - \frac{Q_K}{P_{KV}} C_K(t), \tag{6-5}$$

Liver

$$V_L \frac{\partial}{\partial t} C_L(t) = Q_L C_A(t) - \left(\frac{Q_L}{P_{LV}} + \frac{V_{MAX}}{K_M P_{LV}} + K_B \right) C_L(t) + f_{abs}^{ORAL} K_{ORAL} A_{ORAL}(t), \tag{6-6}$$

Blood

$$f_{AB} V_B \frac{\partial}{\partial t} C_A(t) = Q_C C_{VLU}(t) - Q_C C_A(t), \tag{6-7}$$

$$(1 - f_{AB}) V_B \frac{\partial}{\partial t} C_V(t) = C_{IV}(t) - Q_C C_V(t) + \frac{Q_{AT}}{P_{ATV}} C_{AT}(t) + \frac{Q_S}{P_{SV}} C_S(t) + \frac{Q_K}{P_{KV}} C_K(t) +$$

$$\frac{Q_L}{P_{LV}} C_L(t) + \frac{Q_R}{P_{RV}} C_R(t), \tag{6-8}$$

Rest of the body

$$V_R \frac{\partial}{\partial t} C_R(t) = Q_R C_A(t) - \frac{Q_R}{P_{RV}} C_R(t), \tag{6-9}$$

Gastrointestinal tract

$$\frac{\partial}{\partial t} A_{GI}(t) = K_B C_L(t) - K_F A_{GI}(t), \tag{6-10}$$

Faeces

$$\frac{\partial}{\partial t} A_F(t) = K_F A_{GI}(t). \tag{6-11}$$

6.7.2 PBPK modeling of 3-OHBaP kinetics

Lungs

$$f_{LU} V_{LU} \frac{\partial}{\partial t} C_{vlu}(t) = Q_C C_v(t) - Q_C C_{vlu}(t) - PA_{lu} \left[C_{vlu}(t) - \frac{C_{lu}(t)}{P_{lua}} \right], \tag{6-12}$$

$$(1 - f_{LU}) V_{LU} \frac{\partial}{\partial t} C_{lu}(t) = PA_{lu} \left[C_{vlu}(t) - \frac{C_{lu}(t)}{P_{lua}} \right], \tag{6-13}$$

Adipose tissues

$$f_{AT}V_{AT}\frac{\partial}{\partial t}C_{vat}(t) = Q_{AT}C_a(t) - Q_{AT}C_{vat}(t) - PA_{at}\left[C_{vat}(t) - \frac{C_{at}(t)}{P_{atv}}\right], \tag{6-14}$$

$$(1 - f_{AT})V_{AT}\frac{\partial}{\partial t}C_{at}(t) = PA_{at}\left[C_{vat}(t) - \frac{C_{at}(t)}{P_{atv}}\right], \tag{6-15}$$

Skin

$$V_S\frac{\partial}{\partial t}C_s(t) = Q_SC_a(t) - \frac{Q_S}{P_{sv}}C_s(t), \tag{6-16}$$

Kidneys

$$f_KV_K\frac{\partial}{\partial t}C_{vk}(t) = Q_KC_a(t) - (Q_K + K_{kbr})C_{vk}(t) - PA_k\left[C_{vk}(t) - \frac{C_k(t)}{P_{kv}}\right], \tag{6-17}$$

$$(1 - f_K)V_K\frac{\partial}{\partial t}C_k(t) = PA_k\left[C_{vk}(t) - \frac{C_k(t)}{P_{kv}}\right], \tag{6-18}$$

Liver

$$V_L\frac{\partial}{\partial t}C_l(t) = Q_LC_a(t) - \left(\frac{Q_l}{P_{lv}} + \frac{V_{max}}{K_mP_{lv}} + K_b\right)C_l(t) + K_{gil}A_{gi}(t) + f_{3OHBAP}\frac{V_{MAX}}{K_MP_{LV}}C_L(t), \tag{6-19}$$

Blood

$$f_{AB}V_B\frac{\partial}{\partial t}C_a(t) = Q_CC_{vlu}(t) - Q_CC_a(t), \tag{6-20}$$

$$(1 - f_{AB})V_B\frac{\partial}{\partial t}C_v(t) = -Q_CC_v(t) + Q_{AT}C_{vat}(t) + \frac{Q_S}{P_{sv}}C_s(t) + Q_KC_{vk}(t) + \frac{Q_L}{P_{lv}}C_l(t) +$$

$$\frac{Q_R}{P_{rv}}C_r(t), \tag{6-21}$$

Rest of the body

$$V_R\frac{\partial}{\partial t}C_r(t) = Q_RC_a(t) - \frac{Q_R}{P_{rv}}C_r(t), \tag{6-22}$$

Gastrointestinal tract

$$\frac{\partial}{\partial t}A_{gi}(t) = K_bC_l(t) - \left(K_f + K_{gil}\right)A_{gi}(t), \tag{6-23}$$

Faeces

$$\frac{\partial}{\partial t}A_f(t) = K_fA_{gi}(t), \tag{6-24}$$

Bladder

$$\frac{\partial}{\partial t}A_{br}(t) = K_{kbr}C_{vk}(t) - K_uA_{br}(t), \tag{6-25}$$

Urine

$$\frac{\partial}{\partial t} A_u(t) = K_u A_{br}(t).$$ (6-26)

6.7.3 Toxicokinetic modeling of 3-OHBaP

$$\frac{\partial}{\partial t} A_{iv}(t) = -ka_{iv} A_{iv}(t) + D_{iv}(t),$$ (6-27)

$$\frac{\partial}{\partial t} A_{der}(t) = -ka_{der} A_{der}(t) + D_{der}(t),$$ (6-28)

$$\frac{\partial}{\partial t} A_{ora}(t) = -ka_{oral} A_{oral}(t) + D_{oral}(t),$$ (6-29)

$$\frac{\partial}{\partial t} A_{inh}(t) = -ka_{inh} A_{inh}(t) + D_{inh}(t),$$ (6-30)

$$\frac{\partial}{\partial t} B(t) = -k_b B(t) + ka_{iv} A_{iv}(t) + ka_{der} A_{der}(t) + ka_{oral} A_{oral}(t) + ka_{inh} A_{inh}(t),$$ (6-31)

$$\frac{\partial}{\partial t} U_{inh}(t) = \alpha k_b B(t).$$ (6-32)

6.8 References

1. ATSDR (1995) Toxicological Profile for Polycyclic Aromatic Hydrocarbons. Agency for Toxic Substances and Disease Registry.

2. Scientific Committee on Food (2002) Polycyclic Aromatic Hydrocarbons – Occurrence in foods, dietary exposure and health effects. Directorate C - Scientific Opinions: EUROPEAN COMMISSION.

3. Gendre C, Lafontaine M, Delsaut P, Simon P (2004) Exposure to polycyclic aromatic hydrocarbons and excretion of urinary 3-hydroxybenzo[A]pyrene: Assessment of an appropriate sampling time. Polycyclic Aromatic Compounds 24: 433-439.

4. Lafontaine M, Gendre C, Delsaut P, Simon P (2004) Urinary 3-hydroxybenzo[A]pyrene as a biomarker of exposure to polycyclic aromatic hydrocarbons: An approach for determining a biological limit value. Polycyclic Aromatic Compounds 24: 441-450.

5. Lafontaine M, Gendre C, Morele Y, Laffitte-Rigaud G (2002) Excretion of urinary 1-hydroxypyrene in relation to the penetration routes of polycyclic aromatic hydrocarbons. Polycyclic Aromatic Compounds 22: 579-588.

6. IARC (2010) IARC monographs on the evaluation of carcinogenic risks to humans. Ingested nitrate and nitrite, and cyanobacterial peptide toxins. IARC monographs on the evaluation of carcinogenic risks to humans / World Health Organization, International Agency for Research on Cancer 94: v-vii, 1-412.

7. Scheepers PT, van Houtum J, Anzion RB, Champmartin C, Hertsenberg S, et al. (2009) The occupational exposure of dermatology nurses to polycyclic aromatic hydrocarbons - evaluating the effectiveness of better skin protection. Scandinavian journal of work, environment & health 35: 212-221.

8. Jongeneelen FJ, Anzion RBM, Henderson PT (1987) Determination of Hydroxylated Metabolites of Polycyclic Aromatic-Hydrocarbons in Urine. Journal of Chromatography-Biomedical Applications 413: 227-232.

9. Jongeneelen FJ, Bos RP, Anzion RBM, Theuws JLG, Henderson PT (1986) Biological Monitoring of Polycyclic Aromatic-Hydrocarbons - Metabolites in Urine. Scandinavian Journal of Work Environment & Health 12: 137-143.

10. Forster K, Preuss R, Rossbach B, Bruning T, Angerer J, et al. (2008) 3-Hydroxybenzo[a]pyrene in the urine of workers with occupational exposure to polycyclic aromatic hydrocarbons in different industries. Occupational and environmental medicine 65: 224-229.

11. Gundel J, Schaller KH, Angerer J (2000) Occupational exposure to polycyclic aromatic hydrocarbons in a fireproof stone producing plant: biological monitoring of 1-hydroxypyrene, 1-, 2-, 3- and 4-hydroxyphenanthrene, 3-hydroxybenz(a)anthracene and 3-hydroxybenzo(a)pyrene. International archives of occupational and environmental health 73: 270-274.

12. Likhachev AJ, Beniashvili D, Bykov VJ, Dikun PP, Tyndyk ML, et al. (1992) Biomarkers for individual susceptibility to carcinogenic agents: excretion and carcinogenic risk of benzo[a]pyrene metabolites. Environmental health perspectives 98: 211-214.

13. Payan JP, Lafontaine M, Simon P, Marquet F, Champmartin-Gendre C, et al. (2009) 3-Hydroxybenzo(a)pyrene as a biomarker of dermal exposure to benzo(a)pyrene. Archives of toxicology 83: 873-883.

14. Rey-Salgueiro L, Garcia-Falcon MS, Martinez-Carballo E, Gonzalez-Barreiro C, Simal-Gandara J (2008) The use of manures for detection and quantification of polycyclic aromatic hydrocarbons and 3-hydroxybenzo[a] pyrene in animal husbandry. Science of the Total Environment 406: 279-286.

15. Campo L, Rossella F, Pavanello S, Mielzynska D, Siwinska E, et al. (2010) Urinary profiles to assess polycyclic aromatic hydrocarbons exposure in coke-oven workers. Toxicology letters 192: 72-78.

16. Bouchard M, Brunet RC, Droz PO, Carrier G (2001) A biologically based dynamic model for predicting the disposition of methanol and its metabolites in animals and humans. Toxicol Sci 64: 169-184.

17. Bouchard M, Carrier G, Brunet RC (2008) Assessment of absorbed doses of carbaryl and associated health risks in a group of horticultural greenhouse workers. Int Arch Occup Environ Health 81: 355-370.

18. Bouchard M, Carrier G, Brunet RC, Dumas P, Noisel N (2006) Biological monitoring of exposure to organophosphorus insecticides in a group of horticultural greenhouse workers. Ann Occup Hyg 50: 505-515.

19. Bouchard M, Gosselin NH, Brunet RC, Samuel O, Dumoulin MJ, et al. (2003) A toxicokinetic model of malathion and its metabolites as a tool to assess human exposure and risk through measurements of urinary biomarkers. Toxicological sciences : an official journal of the Society of Toxicology 73: 182-194.

20. Heredia-Ortiz R, Bouchard M, Marie-Desvergne C, Viau C, Maitre A (2011) Modeling of the internal kinetics of benzo(a)pyrene and 3-hydroxybenzo(a)pyrene biomarker from rat data. Toxicological sciences : an official journal of the Society of Toxicology 122: 275-287.

21. Gerlowski LE, Jain RK (1983) Physiologically based pharmacokinetic modeling: principles and applications. Journal of pharmaceutical sciences 72: 1103-1127.

22. Lipscomb JC, Haddad S, Poet T, Krishnan K (2012) Physiologically-based pharmacokinetic (PBPK) models in toxicity testing and risk assessment. Adv Exp Med Biol 745: 76-95.

23. Peters SA (2011) Physiologically based pharmacokinetic (PBPK) modeling and simulations : principles, methods, and applications in the pharmaceutical industry. Hoboken, N.J.: Wiley. xvii, 430 p. p.

24. Sangster J (1989) Octanol-Water Partition-Coefficients of Simple Organic-Compounds. Journal of Physical and Chemical Reference Data 18: 1111-1229.

25. Peyret T, Poulin P, Krishnan K (2010) A unified algorithm for predicting partition coefficients for PBPK modeling of drugs and environmental chemicals. Toxicol Appl Pharmacol 249: 197-207.

26. Schmitt W (2008) General approach for the calculation of tissue to plasma partition coefficients. Toxicol In Vitro 22: 457-467.

27. Heredia-Ortiz R, Bouchard M (2013) Understanding the linked kinetics of benzo(a)pyrene and 3-hydroxybenzo(a)pyrene biomarker of exposure using physiologically-based pharmacokinetic modelling in rats. Submitted to Journal of Pharmacokinetics and Pharmacodynamics.

28. Brown RP, Delp MD, Lindstedt SL, Rhomberg LR, Beliles RP (1997) Physiological parameter values for physiologically based pharmacokinetic models. Toxicology and Industrial Health 13: 407-484.

29. Davies B, Morris T (1993) Physiological-Parameters in Laboratory-Animals and Humans. Pharmaceutical Research 10: 1093-1095.

30. Bevington PR, Robinson DK (2003) Data reduction and error analysis for the physical sciences. Boston: McGraw-Hill. xi, 320 p. p.

31. Morimoto Y, Hatanaka T, Sugibayashi K, Omiya H (1992) Prediction of Skin Permeability of Drugs - Comparison of Human and Hairless Rat Skin. Journal of Pharmacy and Pharmacology 44: 634-639.

32. Saltelli A (2004) Sensitivity analysis in practice a guide to assessing scientific models. Chichester ; Hoboken, NJ: Wiley,. pp. xi, 219 p.

33. Saltelli A, Wiley InterScience (Online service) (2008) Global sensitivity analysis the primer. Chichester, England ; Hoboken, NJ: John Wiley,. pp. x, 292 p.

34. Cao D, Yoon CH, Shin BS, Kim CH, Park ES, et al. (2005) Effects of aloe, aloesin, or propolis on the pharmacokinetics of benzo[a]pyrene and 3-OH-benzo[a]pyrene in rats. Journal of toxicology and environmental health Part A 68: 2227-2238.

35. Marie C, Bouchard M, Heredia-Ortiz R, Viau C, Maitre A (2010) A toxicokinetic study to elucidate 3-hydroxybenzo(a)pyrene atypical urinary excretion profile following

intravenous injection of benzo(a)pyrene in rats. Journal of applied toxicology : JAT 30: 402-410.

36. Weyand EH, Bevan DR (1986) Benzo(a)pyrene disposition and metabolism in rats following intratracheal instillation. Cancer research 46: 5655-5661.

37. Lafontaine M, Payan JP, Delsaut P, Morele Y (2000) Polycyclic aromatic hydrocarbon exposure in an artificial shooting target factory: assessment of 1-hydroxypyrene urinary excretion as a biological indicator of exposure. Ann Occup Hyg 44: 89-100.

38. Simon P, Lafontaine M, Delsaut P, Morele Y, Nicot T (2000) Trace determination of urinary 3-hydroxybenzo[a]pyrene by automated column-switching high-performance liquid chromatography. J Chromatogr B Biomed Sci Appl 748: 337-348.

39. Barbeau D, Maitre A, Marques M (2011) Highly sensitive routine method for urinary 3-hydroxybenzo[a]pyrene quantitation using liquid chromatography-fluorescence detection and automated off-line solid phase extraction. The Analyst 136: 1183-1191.

40. Bouchard M, Viau C (1996) Urinary excretion kinetics of pyrene and benzo(a)pyrene metabolites following intravenous administration of the parent compounds or the metabolites. Toxicology and applied pharmacology 139: 301-309.

41. Jongeneelen, F. and W. ten Berge, Simulation of urinary excretion of 1-hydroxypyrene in various scenarios of exposure to polycyclic aromatic hydrocarbons with a generic, cross-chemical predictive PBTK-model. International Archives of Occupational and Environmental Health, 2012. 85(6): p. 689-702.

42. Sobus, J.R., et al., Comparing Urinary Biomarkers of Airborne and Dermal Exposure to Polycyclic Aromatic Compounds in Asphalt-Exposed Workers. Annals of Occupational Hygiene, 2009. 53(6): p. 561-571.

43. Cavallari, J.M., et al., Predictors of Dermal Exposures to Polycyclic Aromatic Compounds Among Hot-Mix Asphalt Paving Workers. Annals of Occupational Hygiene, 2012. 56(2): p. 125-137.

44. Kriech, A.J., et al., Study Design and Methods to Investigate Inhalation and Dermal Exposure to Polycyclic Aromatic Compounds and Urinary Metabolites from Asphalt

Paving Workers: Research Conducted through Partnership. Polycyclic Aromatic Compounds, 2011. 31(4): p. 243-269.

45. McClean, M.D., et al., Inhalation and dermal exposure among asphalt paving workers. Annals of Occupational Hygiene, 2004. 48(8): p. 663-671.

46. Osborn, L.V., et al., Pilot Study for the Investigation of Personal Breathing Zone and Dermal Exposure Using Levels of Polycyclic Aromatic Compounds (PAC) and PAC Metabolites in the Urine of Hot-Mix Asphalt Paving Workers. Polycyclic Aromatic Compounds, 2011. 31(4): p. 173-200.

6.9 Figures Legends

Figure 6-1: PBPK model of the kinetics of BaP and 3-OHBaP in humans.

Figure 6-2: Single compartment model of the kinetics of BaP and 3-OHBaP in humans.

Figure 6-3: Model simulations of data from an artificial shooting target factory. Comparison of model simulations (lines) with observed data on the time courses of 3-OHBaP in the urine of subjects exposed to PAHs in an artificial shooting target factory (triangles - left-axis). The light gray bars (right-axis) indicate the simulated BaP inhalation exposure scenarios (concentration and time), the white bars (right-axis) indicate the simulated BaP dermal exposure scenarios (concentration and time) while the black bars (right-axis) show the measured inhalation exposure scenarios (measured air concentration (ng/m^3 converted to fmol/mL) and documented time-of-shift; see also Table 6-3). The black solid lines represent PBPK model simulation considering an exposure by the dermal route solely while the dark gray solid lines represent a simulated inhalation. The black dotted lines represent toxicokinetic model simulation considering a dermal exposure solely while the dark gray dotted lines represent a simulated exposure by inhalation. All inhalation concentrations measured in ng/m^3 were expressed in nmol/m^3 and converted to fmol/mL (multiplied by 10^3 so that all the scenarios could be graphically represented on the same figure for comparison).

Figure 6-4: Model simulations of data from a carbon disk brake production factory. Comparison of model simulations (lines) with observed data on the time courses of 3-OHBaP in the urine of workers exposed to PAHs in a carbon disk brake production plant (triangles - left-axis). The light gray bars (right-axis) indicate the simulated BaP inhalation exposure scenarios (concentration and time), the white bars (right-axis) indicate the simulated BaP dermal exposure scenarios (concentration and time) while the black bars (right-axis) show the measured inhalation exposure scenarios (measured air concentration (ng/m^3 converted to fmol/mL) and documented time-of-shift; see also Table 6-4). The black solid lines represent PBPK model simulation considering an exposure by the dermal route solely while the dark gray solid lines represent a simulated inhalation. The black dotted lines represent toxicokinetic model simulation considering a dermal exposure solely while the dark gray dotted lines represent a simulated exposure by inhalation. All inhalation concentrations measured in ng/m^3 were expressed in nmol/m^3 and converted to fmol/mL (multiplied by 10^3 so that all the scenarios could be graphically represented on the same figure for comparison).

Figure 6-5: Model simulations of data from a silicon production factory. Comparison of model simulations (lines) with observed data on the time courses of 3-OHBaP in the urine of workers exposed to PAHs in a silicon production industry (triangles - left-axis). The light gray bars (right-axis) indicate the simulated BaP inhalation exposure scenarios (concentration and time), the white bars (right-axis) indicate the simulated BaP dermal exposure scenarios (concentration and time) while the black bars (right-axis) show the measured inhalation exposure scenarios (measured air concentration (ng/m^3 converted to fmol/mL) and documented time-of-shift; see also Table 6-5). The black solid lines represent PBPK model simulation considering an exposure by the dermal route solely while the dark gray solid lines represent a simulated inhalation. The black dotted lines represent toxicokinetic model simulation considering a dermal exposure solely while the dark gray dotted lines represent a simulated exposure by inhalation. All inhalation concentrations measured in ng/m^3 were expressed in nmol/m^3 and converted to fmol/mL (multiplied by 10^3 so that all the scenarios could be graphically represented on the same figure for comparison).

198

6.10 Tables

Table 6-1 : Rat-to-human extrapolated key parameter values in the PBPK and toxicokinetic models[a]

Parameters	Units	PBPK model Rats	PBPK model Humans	Toxicokinetic model Rats	Toxicokinetic model Humans
Metabolic constants	h^{-1}	$\dfrac{V_{max}}{K_M}\Big|_{BaP}$ $\dfrac{V_{max}}{K_M}\Big|_{3-OHBa1}$	$C_{rat-human}\dfrac{V_{max}}{K_M}\Big|^b_{BaP}$ $C_{rat-human}\dfrac{V_{max}}{K_M}\Big|^b_{3-OHBaP}$	k_b	$\left(\dfrac{V_{AT}^{rat}}{Q_{AT}^{rat}}\right)\left(\dfrac{Q_{AT}^{hum}}{V_{AT}^{hum}}\right)\left(\dfrac{125}{1.31}\right)^c C_{rat-human}{}^b k_b$
Glomerular filtration rate	$mL\ h^{-1}$	K_{khr}	$\dfrac{125^c}{1.31} K_{khr}$		
Bile flow rate	$mL\ h^{-1}$	K_B	$\dfrac{350^d}{22.5} K_B$	-	-
Skin permeability coefficient	$cm\ h^{-1}$	k_P	$f_s^e k_P$	$D_{dermal}(t)$	$f_s^e D_{dermal}(t)$

[a] All the parameters are defined in the Appendix with the usual nomenclature: V_{max} representing the maximum velocity rate of metabolism, K_m the Michaelis-Menten affinity rate constant, V_{AT} the adipose tissue volume and Q_{AT} the blood flow rate to adipose tissues.

[b] This is the scaling constant ($C_{rat-human}$ = human biotransformation rate/rat biotransformation rate = 1020.03) for rat-to-human extrapolation of metabolic constants.

[c] This is the ratio of glomerular filtration rates in rats (1.31 mL/min) and in humans (125 mL/min) as reported by Davies and Morris [29].

[d] Values for the bile flow rate in rats (22.5 mL/day) and in humans (350 mL/day) from Davies and Morris [29].

[e] According to the scaling proposed by Morimoto et al. [31] relating the permeability coefficient in humans and hairless rat skin, and octanol-water partition coefficient, $f_s = \left(\frac{1.17 \times 10^{-7}(6.19^{0.751}) + 2.73 \times 10^{-8}}{14.78 \times 10^{-7}(6.19^{0.589}) + 8.33 \times 10^{-8}} \right)$.

Table 6-2 : Human model parameters and sensitivity results.

		BaP		3-OHBaP	
Matrix	**Parameter**	**Value**	**Sensitivity**[a]	**Value**	**Sensitivity**[a]
Partition coefficients					
Lungs	P_{LUA} and P_{lua}	2670.00	±39.6%	2.92	±39.6%
Adipose tissues	P_{ATV} and P_{atv}	65.90	±39.6%	1.42	±39.6%
Skin	P_{SV} and P_{sv}	1.87	±39.6%	0.80	±39.6%
Kidneys	P_{KV} and P_{kv}	2.08	±39.6%	40.40	±39.6%
Liver	P_{LV} and P_{lv}	12.90	±39.6%	1.83	±39.6%
Rest of the body	P_{RV} and P_{rv}	10.00	±29.6%	1.00	±39.6%
Permeability coefficients					
Lungs	PA_{LU} and PA_{lu} [mL/h]	80.70	±39.6%	0.20	±39.6%
Adipose tissues	PA_{at} [mL/h]	-	-	0.711	±39.6%
Kidneys	PA_k [mL/h]	-	-	12.90	±39.6%
Metabolic Constants					
Total metabolites	$\sum_{Vt} \frac{V_{max}^{(t)}}{K_m^{(t)}}$ [mL/h]	951.69x10³	±38%	37.13x10³	±4.8%
Fraction of 3-OHBaP	$\left(\frac{V_{max}^{(3-OHBaP)}}{K_m^{(3-OHBaP)}}\right)\left(\sum_{Vt}\frac{V_{max}^{(t)}}{K_m^{(t)}}\right)^{-1}$	0.185	±4.0%	-	-

	Matrix	**Parameter**	**BaP**		**3-OHBaP**	
			Value	**Sensitivity[a]**	**Value**	**Sensitivity[a]**
Elimination rates	Biliary	K_B and K_b [1/h]	0.338	±39.6%	663.80	±39.6%
	Urinary	K_{Lb} [1/h]	-	-	60.40	±4.6%
		K_{bu} [1/h]	-	-	0.102	±12%
	Faecal	K_F and K_f [1/h]	0.334	±39.6%	0.173	±39.6%
		K_{gil} [1/h]	-	-	0.00693	±39.6%
Absorption constants	Dermal	k_P [cm/h]	0.00132	±39.6%	-	-
		P_{DV}	1.0	±39.6%	-	-
	Inhalation	P_B	2.04	±39.6%	-	-

[a] Range of parameter variation during Monte Carlo simulation to obtain 90.51 ± 1.15% of runs (n~O(10³)) within a maximum variation of ±10% in the simulated urinary excretion profiles compared to default parameter values.

Table 6-3 : Simulated dermal and inhalation exposure scenarios compared with measured BaP inhalation exposure scenario (air concentrations and time-of-shifts) in subjects exposed to PAHs in an artificial shooting target factory (observed data from [37]).

Subject[a]	Route-of-entry	Measured exposure scenarios		Simulated exposure scenarios	
		Day 1	Day 2	Day 1	Day 2
1	Inhalation	233 ng/m³ 12:00-21:00	302 ng/m³ 12:00-21:00	2749.4 ng/m³ 12:00-21:00	3322 ng/m³ 12:00-21:00
	Dermal exposure			19.8 fmol/mL 12:00-21:00	26.1 fmol/mL 12:00-21:00
2	Inhalation	208 ng/m³ 12:00-24:00	285 ng/m³ 12:00-21:00	1040 ng/m³ 12:00-21:00	1425 ng/m³ 13:00-1:00
	Dermal exposure			7.5 fmol/mL 12:00-12:00	10.8 fmol/mL 12:00-21:00
3	Inhalation	537 ng/m³ 13:00-21:00	991 ng/m³ 13:00-21:00	13962 ng/m³ 13:00-21:00	21802 ng/m³ 13:00-21:00
	Dermal exposure			47.1 fmol/mL 13:00-21:00	123.6 fmol/mL 8:00-20:00
4[b]	Inhalation	422 ng/m³ 13:00-21:00	583 ng/m³ 13:00-21:00	6330 ng/m³ 13:00-10:00	8745 ng/m³ 13:00-21:00
	Dermal exposure			36.2 fmol/mL 20:00-10:00	39.5 fmol/mL 20:00-13:00
5[c]	Inhalation	519 ng/m³ 13:00-21:00	684 ng/m³ 13:00-21:00	7785 ng/m³ 13:00-21:00	8208 ng/m³ 13:00-21:00
	Dermal exposure			36.2 fmol/mL 13:00-10:00	56.8 fmol/mL 20:00-16:00

[a] Subjects 1 and 2 were the industrial hygienists on site for the biomonitoring of exposure to PAH in workers. None of the workers wore respiratory protection but two of the workers wore gloves regularly.

[b] This worker was used to fit the human metabolism rate of BaP and 3-OHBaP.

[c] Worker 5 not presented in Figure 6-3.

Table 6–4 : Simulated dermal and inhalation exposure scenarios compared with measured BaP inhalation exposure scenario (air concentrations and time-of-shifts) in workers exposed to PAHs in a carbon disk brake production plant (observed data from [3]).

Worker[a]	Route-of-entry	Measured exposure scenarios		Simulated exposure scenarios	
		Day 1	Day 2	Day 1	Day 2
6[b]	Inhalation	8 ng/m³ 8:00-16:00		240 ng/m³ 8:00-16:00	240 ng/m³ 8:00-16:00
	Dermal exposure			2 fmol/mL 8:00-16:00	4.6 fmol/mL 8:00-16:00
7	Inhalation	9300 ng/m³ 8:00-16:00	560 ng/m³ 8:00-16:00	4650 ng/m³ 8:00-16:00	5600 ng/m³ 8:00-16:00
	Dermal exposure			32.6 fmol/mL 8:00-8:00	106.3 fmol/mL 8:00-16:00
8	Inhalation	5650 ng/m³ 8:00-16:00	270 ng/m³ 8:00-16:00	6215 ng/m³ 8:00-16:00	4050 ng/m³ 8:00-16:00
	Dermal exposure			43.5 fmol/mL 8:00-24:00	21.9 fmol/mL 8:00-16:00
9	Inhalation	1500 ng/m³ 8:00-16:00	65 ng/m³ 8:00-16:00	7500 ng/m³ 8:00-16:00	6500 ng/m³ 8:00-16:00
	Dermal exposure			41 fmol/mL 8:00-16:00	123.4 fmol/mL 8:00-16:00
10	Inhalation	775 ng/m³ 8:00-16:00	63 ng/m³ 8:00-16:00	2712.5 ng/m³ 8:00-8:00	5670 ng/m³ 8:00-16:00

Worker[a]	Route-of-entry	Measured exposure scenarios		Simulated exposure scenarios	
		Day 1	Day 2	Day 1	Day 2
	Dermal exposure			16.6 fmol/mL 8:00-8:00	351.6 fmol/mL 8:00-9:00

[a] Worker 7 wore a paper mask; worker 6 and 9 did not wear any respiratory protection equipment; worker 10 wore a cartridge mask.

[b] Worker 6 not presented in Figure 6-4.

207

Table 6-5 : Simulated dermal and inhalation exposure scenarios compared with measured BaP inhalation exposure scenario (air concentrations and time-of-shifts) in workers exposed to PAHs during a metallurgical furnace repair in a silicon production plant.

Worker	Units	Measured exposure scenarios				Simulated exposure scenarios			
		Day 1	Day 2	Day 3	Day 4	Day 1	Day 2	Day 3	Day 4
11 [a]	Inhalation	3619 ng/m³ [b] 6:00-14:00	ND [b,c]	ND [b,c]	18.5 ng/m³ 6:00-14:00	6514.2 ng/m³ 6:00-14:00	1881.8 ng/m³ 6:00-14:00	1737.1 ng/m³ 6:00-14:00	3619 ng/m³ 6:00-14:00
	Dermal exposure	ND [b,c]	ND [b,c]		33.3 fmol/mL 6:00-21:00	23.6 fmol/mL 6:00-17:00	15.1 fmol/mL 6:00-14:00	22.6 fmol/mL 6:00-14:00	
12 [a]	Inhalation	2668.1 ng/m³ [b] 6:00-14:00	ND [b,c]	ND [b,c]	77 ng/m³ 6:00-14:00	8537.9 ng/m³ 6:00-21:00	800.4 ng/m³ 6:00-17:00	3201.7 ng/m³ 6:00-14:00	4002.1 ng/m³ 6:00-22:00
	Dermal exposure				49.9 fmol/mL 6:00-18:00	10 fmol/mL 6:00-5:00	9.1 fmol/mL 6:00-5:00	10.8 fmol/mL 6:00-2:00	
13 [a]	Inhalation	2121.5 ng/m³ [b] 6:00-14:00	ND [b,c]	ND [b,c]	28 ng/m³ 6:00-14:00	2121.5 ng/m³ 6:00-14:00	2651.9 ng/m³ 6:00-14:00	424.3 ng/m³ 6:00-5:00	28 ng/m³ 6:00-14:00
	Dermal exposure				21.5 fmol/mL 6:00-14:00	29 fmol/mL 6:00-14:00	7.8 fmol/mL 6:00-14:00	0.3 fmol/mL 6:00-14:00	
14 [a]	Inhalation	3996.7 ng/m³ [b] 6:00-14:00	ND [b,c]	ND [b,c]	51 ng/m³ 6:00-14:00	45962 ng/m³ 6:00-14:00	399.7 ng/m³ 6:00-14:00	3996.7 ng/m³ 6:00-14:00	2805 ng/m³ 6:00-14:00
	Dermal exposure				498.6 fmol/mL 6:00-14:00	0 fmol/mL 6:00-14:00	52 fmol/mL 6:00-18:00	13.3 fmol/mL 6:00-6:00	

[a] Workers were exposed during 4 consecutive days (Tuesday to Friday). No collective protection equipment was in place, but workers wore masks (with type ABEK2P2 or A2P3 cartridges) and leather handling gloves.

[b] Days 1 to 4 correspond to the four exposure days, hence days where workers were performing repair tasks exposing them to PAHs. However, air concentrations were measured on the first exposure day of the week (Tuesday) by the team of Professor Maitre and taken to be equal on days 2 and 3 of exposure (Wednesday, Thursday). Air concentrations were also monitored on the last exposure day (Friday; day 4 of exposure).

[c] Not determined.

6.11 Figures

Figure 6-1: PBPK model of the kinetics of BaP and 3-OHBaP in humans.

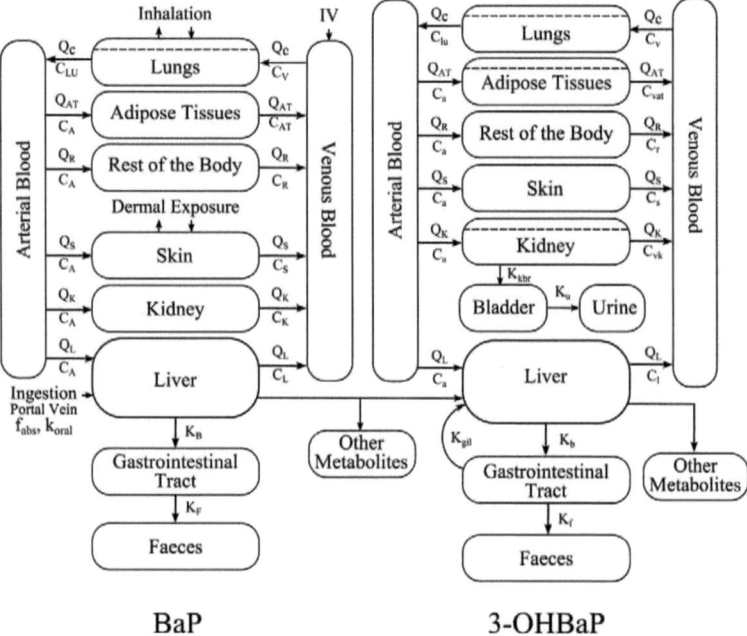

Figure 6-2: Single compartment model of the kinetics of BaP and 3-OHBaP in humans.

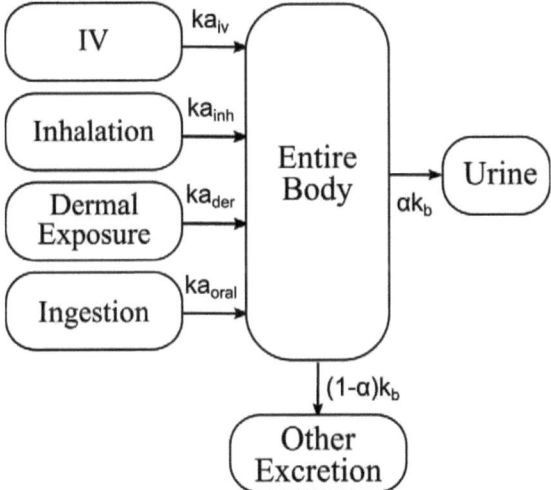

Figure 6-3: Model simulations of data from an artificial shooting target factory.

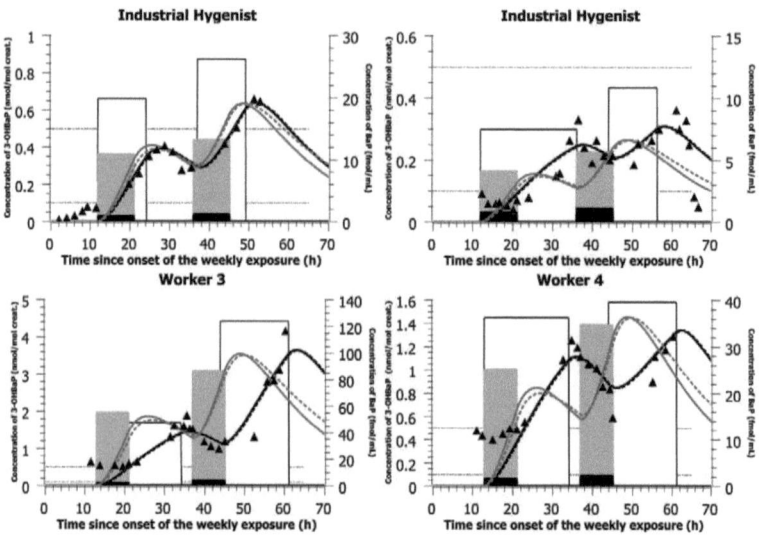

Figure 6-4: Model simulations of data from a carbon disk brake production factory.

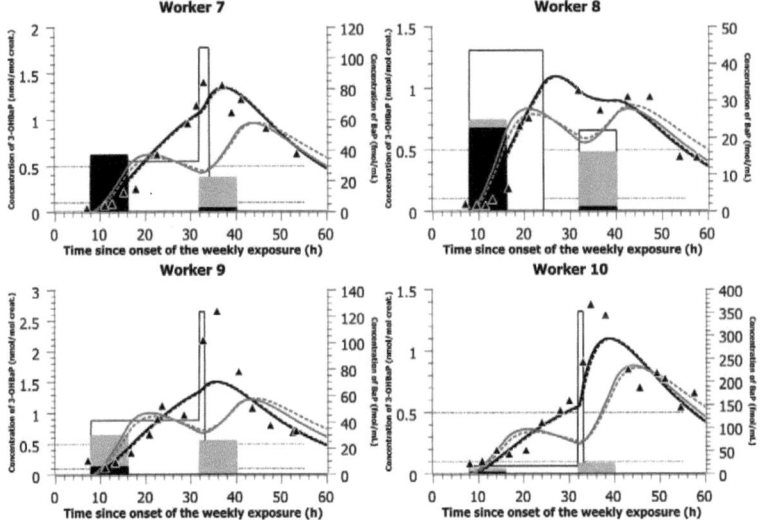

Figure 6-5: Model simulations of data from a silicon production factory.

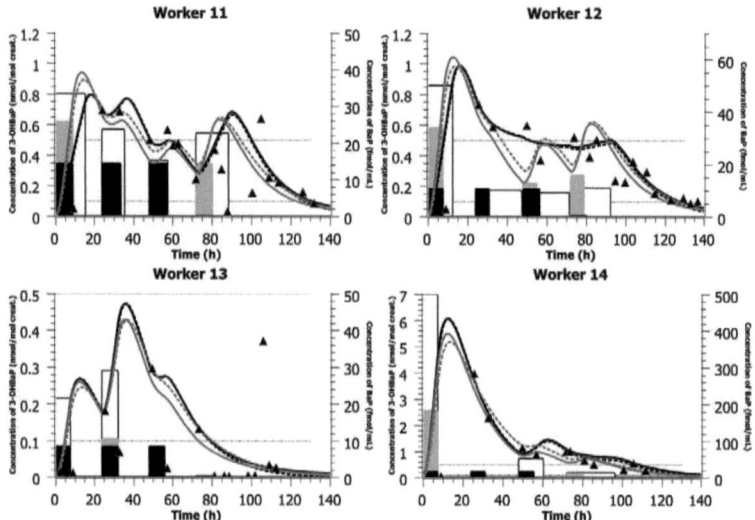

7 Discussion générale et conclusion

7.1 Les modèles cinétiques développés

7.1.1 Les modèles à compartiments

Deux types de modèles à compartiments ont été développés pour représenter la cinétique du BaP et de son métabolite 3-OHBaP lors de l'exposition de rats au BaP par quatre voies d'administration différentes : intraveineuse, respiratoire, cutanée et orale. Nous avons démontré la validité de ces modèles à décrire la cinétique du BaP chez les rats et les humains. En effet, ces modèles ont permis de simuler plusieurs profils sanguins, tissulaires et urinaires obtenus à partir de différentes études expérimentales *in vivo*, dans diverses conditions d'exposition.

Le premier modèle a été élaboré pour décrire la cinétique du BaP et du 3-OHBaP dans l'organisme (sang, le plus grand nombre de tissus et excrétas), sur la base des données *in vivo* animales disponibles dans la littérature scientifique. Cette approche nous a permis de démontrer que :

i.   La modélisation suivant une cinétique de premier ordre est adéquate pour décrire les profils temporels du BaP et celle du 3-OHBaP chez les rats, lorsqu'ils sont exposés à une dose de BaP variant entre 1,7 nmol et 9,9 µmol;

ii. La modélisation à un compartiment est adéquate pour simuler des profils temporels dans des matrices biologiques en équilibre dynamique avec le sang. Tel a été le cas de la concentration sanguine du BaP qui est en équilibre cinétique avec les concentrations observées dans le foie, les reins et la peau. De plus, la concentration sanguine du 3-OHBaP est en équilibre avec les concentrations observées dans le foie et la peau;

iii. La résolution analytique des équations différentielles facilite la détermination des paramètres des modèles;

iv. Un modèle cinétique adapté à l'humain peut être développé en extrapolant les paramètres cinétiques clés (taux d'élimination des poumons et des tissus adipeux) de l'animal à l'humain;

v. La modélisation toxicocinétique à compartiments permet de reproduire les profils urinaires du 3-OHBaP chez des travailleurs exposés au BaP, par l'extrapolation d'un modèle cinétique construit à partir des études expérimentales menées sur des rats.

Le deuxième modèle cinétique a été construit en utilisant un compartiment unique pour décrire la cinétique d'élimination du 3-OHBaP suite à l'exposition au BaP par quatre voies d'administration différentes : intraveineuse, respiratoire, cutanée et orale. Cette modélisation nous a permis de démontrer que les mêmes résultats, listés auparavant, peuvent être obtenus avec l'utilisation d'un modèle simple. Quand il s'agit de déterminer la relation cinétique entre la dose d'exposition au BaP et l'élimination urinaire du 3-OHBaP, la complexité du modèle à compartiments utilisé n'a pas d'importance. Ceci ne s'explique pas uniquement par le fait que les deux mesures utilisées (la concentration du BaP au site d'absorption et celle du 3-OHBaP dans l'urine) suivent une cinétique de premier ordre. En effet, dans le cas où la cinétique présente des mécanismes de saturation, les équations différentielles du modèle à compartiments simple peuvent être modifiées pour tenir compte d'un mécanisme d'ordre supérieur sans avoir recours à l'ajout de nouveaux compartiments. L'explication réside dans le nombre de mesures à modéliser. Un seul compartiment peut tenir compte d'une relation entre deux mesures distinctes (une à l'entrée et l'autre à la sortie).

217

Si nous devions établir une relation entre trois mesures (disons, la concentration du BaP suite à l'exposition et l'élimination de deux métabolites différents), le nombre minimal de compartiments du modèle serait de deux.

7.1.2 Le modèle PCBP

Un modèle PCBP a été élaboré pour décrire la cinétique du BaP et du 3-OHBaP chez des rats. À partir des mêmes profils tissulaires et sanguins utilisés pour construire les modèles à compartiments, la modélisation à base physiologique a considéré : les tissus adipeux, les poumons, le foie, les reins, la peau, l'urine et les fèces. Le modèle PCBP nous a permis de confirmer que :

i. L'utilisation initiale de la voie intraveineuse est une bonne stratégie pour modéliser d'autres voies d'exposition : respiratoire, cutanée et orale;

ii. L'approximation qui considère que le sang tissulaire est proportionnel au sang veineux effluent (la constante de proportionnalité étant le coefficient de partition) est suffisante pour modéliser le profil temporel tissulaire du BaP et de son métabolite 3-OHBaP;

iii. Dans le processus de modélisation PCBP, la considération d'une cinétique de premier ordre est adéquate pour décrire les profils temporels du BaP et du 3-OHBaP chez les rats lorsqu'ils sont exposés à des doses de BaP variant entre 1,7 nmol et 9,9 μmol;

iv. Le modèle PCPB developpé à partir d'études chez le rat peut être adapté à l'humain en ajustant les paramètres clés tout en considérant les coefficients

218

de partition et les coefficients de perméabilité tissulaire comme étant des invariants selon l'espèce;

v. La métabolisation du BaP en 3-OHBaP peut être modélisée en considérant que le 3-OHBaP représente, en tout temps, une fraction des métabolites totaux formés;

vi. L'utilisation de processus probabilistes, comme les algorithmes Monte-Carlo, est très avantageuse pour déterminer numériquement les paramètres des modèles cinétiques;

vi. La modélisation PCBP permet de reproduire les profils urinaires du 3-OHBaP chez des travailleurs exposés au BaP, par l'extrapolation du modèle cinétique construit à partir de profils sanguins, tissulaires et urinaires chez le rat.

Le modèle PCBP élaboré a été validé en reproduisant un ensemble de données expérimentales indépendantes (Bouchard et Viau, 1997; Cao et al., 2005; Chien et Yeh, 2012; Foth et al., 1988; Gendre et al., 2004; Jongeneelen et al., 1985; Lafontaine et al., 2000; Lee et al., 2003; Likhachev et al., 1992; Marie et al., 2010; Moir et al., 1998; Payan et al., 2009; Ramesh et al., 2001a; Ramesh et al., 2002; Ramesh et al., 2001b; Weyand et Bevan, 1986). Les taux d'excrétion urinaire et les constantes de métabolisation se sont avérés être les paramètres les plus sensibles du modèle PCBP. Cela nous est révélé par les analyses de sensibilité réalisées sur l'excrétion urinaire du 3-OHBaP.

7.1.3 L'extrapolation inter-espèces

L'extrapolation du modèle PCBP de l'animal à l'humain est relativement directe. Les valeurs physiologiques du rat sont remplacées par celles de l'humain; il s'agit

219

notamment de remplacer, par les données spécifiques à l'espèce, les valeurs de volumes sanguins, de poids des tissus, de débits sanguins tissulaires, de taux d'absorption, de taux d'élimination (représentant la sécrétion biliaire et le taux d'élimination urinaire), de débits ventilatoires. Toutes ces caractéristiques décrites sont d'ordre physiologique. Dès lors, leur extrapolation est simple puisque ces paramètres ne dépendent pas de la cinétique du BaP, ni de celle de son métabolite. Cependant, ce n'est pas le cas de la biotransformation. Ainsi, pour extrapoler le modèle à la cinétique du BaP chez l'humain, les taux de métabolisation du BaP et du 3-OHBaP ont été déterminés par ajustement de ce paramètre au profil d'excrétion urinaire d'un travailleur exposé au BaP. Cette approche s'est avérée adéquate pour reproduire les divers profils urinaires de travailleurs exposés aux BaP et ayant un taux de métabolisation similaire. La modélisation PCBP a aussi permis de reproduire les différents scénarios d'exposition de ces travailleurs par différentes voies d'exposition.

Puisque le modèle à compartiments (non physiologique) ne contient pas d'information explicite sur les valeurs physiologiques telles que le volume des tissus ou le débit sanguin, il devient nécessaire lors de l'extrapolation à l'humain d'établir la correspondance entre les paramètres cinétiques utilisés dans le modèle et les variables physiologiques associées. Par exemple, puisque le taux d'excrétion total du 3-OHBaP doit être proportionnel au taux de relâchement du BaP par les tissus adipeux, une correction linéaire a été effectuée entre le taux d'excrétion total du 3-OHBaP des rats et celui des humains. En réalité, tous les tissus contribuent à la valeur finale du taux d'excrétion total du 3-OHBaP. Néanmoins, la vitesse de transfert du BaP des tissus adipeux au sang est le facteur qui régit le taux d'excrétion total du 3-OHBaP sur la période évaluée (de l'ordre d'heures). Le modèle à compartiments (non physiologique), tout comme le modèle PCBP, a permis un bon ajustement du modèle aux profils urinaires de 3-OHBaP chez les travailleurs exposés au BaP, pour toutes les voies d'exposition.

7.1.4 La linéarité de la cinétique du BaP et du 3-OHBaP

Les données expérimentales disponibles dans la littérature (voir tableaux 5-1 et 5-2) et utilisées pour le développement et la validation des modèles semblent suivre une cinétique de premier ordre pour toutes les doses et voies d'exposition testées. Selon Bouchard et Viau (1997), l'excrétion urinaire cumulative du 3-OHBaP au cours des soixante heures suivant l'exposition est de l'ordre de 10 nmol suite à une exposition intraveineuse de 10 µmol de BaP. Suite à une exposition par voie intraveineuse de 19,8 µmol de BaP, Lee *et al.* (2003) ont observé une excrétion urinaire cumulative du 3-OHBaP deux fois plus élevée (23,8 nmol) durant les 60 h post-exposition. De la même façon, lors d'une exposition par voie intraveineuse de 0,9 µmol de BaP, Cao *et al.* (2005) ont obtenu approximativement 2,3 nmol de 3-OHBaP dans l'urine 0-60 h post-exposition. Ces résultats suggèrent une relation linéaire entre la dose administrée de BaP (par voie intraveineuse) et les niveaux urinaires de 3-OHBaP chez le rat.

La même linéarité par rapport à la dose devrait être observée pour les autres voies d'exposition à l'exception des cas où il y aurait une saturation des processus d'absorption du BaP. Or, dans les études analysées, ce n'est pas le cas des voies cutanée, respiratoire et orale. Dans une étude par voie cutanée, Payan *et al.* (2009) ont exposé des rats à approximativement 0,3 µmol de BaP. Après quarante-huit heures, ils ont observé une excrétion urinaire cumulative de 3-OHBaP de 0,65 nmol. Dans une autre étude par voie cutanée, Jongeneelen *et al.* (1984) ont observé une excrétion urinaire cumulative d'approximativement 30 nmol de 3-OHBaP au cours des quarante-huit heures suivant l'exposition à 12,5 µmol de BaP. Le rapport entre les deux doses d'exposition de BaP est de l'ordre de 500 et le même rapport entre les excrétions urinaires de 3-OHBaP est conservé.

Dans une étude chez des rats exposés à 1 nmol de BaP par instillation intra-trachéale pendant cinq minutes, Weyand et Bevan (1986) ont obtenu une concentration sanguine de BaP de l'ordre de 0,3 pmol/mL six heures après la fin de l'exposition. Dans une autre étude chez des rats exposés par inhalation à 11,4 µmol de BaP pendant quatre heures, Ramesh *et al.* (2001a) ont rapporté une concentration sanguine de l'ordre de 0,1 nmol/mL quatre heures après la fin de l'exposition. Dans une étude par voie orale, Ramesh *et al.* (2001b) ont exposé des rats à 99,1 µmol de BaP. Après huit heures, ils ont observé approximativement 69 µmol de BaP dans le sang. Dans l'étude de Cao *et al.* (2005), les auteurs ont observé 13,5 µmol de BaP dans le sang après cinq heures d'exposition à 19,8 µmol de BaP. Le ratio entre les doses d'exposition du BaP par voie orale et les quantités de BaP observées dans le sang sont du même ordre.

L'ensemble de ces résultats nous a permis de développer des modèles présentant une cinétique de premier ordre, pour toutes les modélisations présentées. En d'autres termes, la modélisation linéaire s'en trouve validée pour la cinétique du BaP et du 3-OHBaP, chez les rats, dans l'intervalle des doses présenté dans la section précédente.

7.1.5 Comparaison avec d'autres modèles publiés

7.1.5.1 Modèles à compartiments

La totalité des modèles à compartiments publiés pour le BaP ont seulement été utilisés pour estimer la demi-vie d'élimination. Un certain nombre de ces études s'est intéressé à la contamination de systèmes aquatiques au BaP (Heinonen *et al.*, 2000; Schuler *et al.*, 2003; Van Hattum et Montanes, 1999; Leversee *et al.*, 1982). Ces études avaient comme objectif de pouvoir utiliser des animaux pour mieux estimer les concentrations du BaP dans l'eau. Ces études ne sont pas pertinentes pour notre modélisation puisqu'elles n'ont pas considéré la cinétique du BaP dans les organismes employés. D'autres études ont utilisé des animaux, qui n'étaient pas des mammifères, pour estimer la vitesse d'élimination du BaP (James *et al.*, 1995; Seubert et Kennedy, 2000). Par conséquent, les demi-vies rapportées dans ces études étaient très différentes de celles qui pourraient être observées chez les mammifères.

La valeur de demi-vie d'élimination du BaP dans le sang (importante dans la période de douze heures suivant l'exposition au BaP) prédite par notre modèle à compartiments est de 3,3 heures pour le BaP. La valeur de demi-vie du BaP dans les tissus adipeux (importante douze heures après l'exposition au BaP), obtenue par le même modèle, est de 27,2 heures pour le BaP. Ces deux valeurs sont du même ordre de grandeur que celles observées par les publications analysées (Cao *et al.*, 2005; Marie *et al.*, 2010; Moir *et al.*, 1998; Payan *et al.*, 2009; Ramesh *et al.*, 2001a; Ramesh *et al.*, 2001b; Weyand et Bevan, 1986; Wiersma et Roth, 1983). Ce résultat vient valider la structure de notre modèle à plusieurs compartiments ainsi que les valeurs de paramètres cinétiques obtenus lors de la construction du modèle.

Bouchard et Viau (1996) ont calculé une demi-vie d'élimination du 3-OHBaP de 8,1 h à partir des profils urinaires établis suite à une injection intraveineuse de 40 μmol/kg de BaP chez le rat. Lafontaine *et al.* (2004) ont documenté de valeurs correspondantes oscillant entre 3,1 et 16,2 heures pour des travailleurs exposés probablement par inhalation. Plus récemment, Chien et Yeh (2012) ont rapporté des valeurs de demi-vie d'élimination du 3-OHBaP dans l'urine se situant entre 2,5 et 4,3 heures pour de volontaires exposés par ingestion d'un repas cuit au charbon de bois. À l'aide de notre modèle à compartiment unique, nous avons obtenu une valeur de demi-vie d'élimination du 3-OHBaP dans l'urine de 3,5 heures pour les rats, et autour de 18 heures pour les travailleurs. Ces valeurs sont du même ordre de grandeur. Ces résultats confirment la capacité de nos modèles à reproduire la vitesse d'élimination du BaP et du 3-OHBaP de l'organisme.

7.1.5.2 Modèles PCBP

Les deux modèles PCBP, précédant notre modélisation, ont des valeurs paramétriques très différentes, même si les deux études tentent de simuler le même ensemble de données (Wiersma et Roth, 1983). Ceci témoigne de possibles divergences observées lors de la construction de modèles sur la base de données *in vitro*. Lors de l'extrapolation des paramètres *in vitro* à des estimations de valeurs *in vivo*, plusieurs facteurs pouvant affecter la valeur des paramètres du modèle doivent être évalués. Le nombre d'hépatocytes par gramme de foie (en moyenne 137 millions), la quantité d'enzymes microsomales par gramme de foie (32 mg en général) et la liaison du BaP aux protéines sanguines peuvent fortement influencer la clairance intrinsèque et donc la vitesse de la métabolisation du BaP dans un organisme vivant (Knaak *et al.*, 2012; Yoon *et al.*, 2014). Ainsi, même si la structure des modèles publiés par Crowell *et al.* (2011) et Pery *et al.* (2011) est identique, les valeurs de leurs paramètres sont différentes.

Le coefficient de partition pour le BaP dans les tissus adipeux obtenu par Crowell *et al.* (2011) est 496,4 tandis que celui obtenu par Pery *et al.* (2011) est 50. Dans le premier cas, la valeur a été obtenue grâce à l'algorithme pour prédire les coefficients de partition de produits organiques à partir des coefficients de partage entre l'eau et l'octanol (Poulin et Krishnan, 1995; Poulin et Theil, 2000). Dans le deuxième cas, le coefficient de partition a été ajusté à des profils sanguins du tétrachlorodibenzo-p-dioxine, un composé lipophile comparable au BaP (Zeilmaker *et al.*, 1997). La valeur obtenue suite à notre modélisation par ajustement à plusieurs profils tissulaires et sanguins a donné un coefficient de partition similaire à celui obtenu par Pery *et al.* (2011) : 65,9. Pery *et al.* (2011) ont commencé avec un grand coefficient de partition de 650 et, par la suite, ont réduit la valeur jusqu'au meilleur ajustement à 50. Dans notre modélisation, la valeur du coefficient de partition pouvait se trouver entre 0.01 et 1000. Suite à l'ajustement, l'algorithme Monte-Carlo a estimé que la meilleure valeur était de 65,9.

Les coefficients de partition tissu : sang estimés pour les tissus pauvrement perfusés étaient similaires dans les trois cas : 6,99 (Crowell *et al.*, 2011), 3,5 (Pery *et al.*, 2011) et 10 (notre modélisation). Les coefficients de partition tissus : sang pour les tissus richement perfusés estimés étaient aussi comparables dans les trois cas : 13,3 (Crowell *et al.*, 2011), 1,6 (Pery *et al.*, 2011) et 1,87/2,08/12,9 (notre modélisation a tenu compte séparément de la peau, des reins et du foie). Dans l'étude de Crowell, ils ont utilisé l'algorithme de Poulin et Krishnan (1995) pour obtenir les valeurs de coefficients de partition tissu : sang pour les tissus pauvrement perfusés (correspondant aux muscles) et les tissus richement perfusés (en considérant le foie). Dans l'étude de Pery, l'ajustement a été effectué par rapport à deux ensembles différents de données non décrites. Ces résultats indiquent que la cinétique du BaP dans les tissus pauvrement et richement perfusés peut être modélisée en utilisant le coefficient de partage octanol : eau et l'algorithme de Poulin et Krishnan (1995).

Les coefficients de partition dans le foie utilisés étaient similaires dans les trois études: 13,3 (Crowell *et al.*, 2011), 25 (Pery *et al.*, 2011) et 12,9 pour notre modélisation. Ce résultat indique que le coefficient de partition foie : sang du BaP peut être estimé à partir du ratio de concentrations du BaP dans l'octanol et dans l'eau. Il n'est pas possible de comparer directement le taux de métabolisation du BaP dans notre modèle avec les autres valeurs rapportées dans la littérature, puisque notre modélisation a supposé un taux de biotransformation linéaire, tandis que les autres modèles l'ont considéré non-linéaire. Cependant, les ratios entre la vitesse maximale de métabolisation et la constante de Michaelis-Menten peuvent être employés pour cette fin. Ainsi, le plus petit ratio, 487,6 mL/h, a été calculé par Crowell *et al.* (2011) à partir de données expérimentales sur la clairance hépatique obtenues *in vitro* (Wiersma et Roth, 1983). À partir de la même étude,

225

Pery *et al.* (2011) ont estimé le ratio à 3755,4 mL/h. Le ratio utilisé par notre modèle a été 933 mL/h. D'une part, ces résultats nous montrent la variabilité de ces paramètres et d'autre part, ils nous indiquent une métabolisation rapide du BaP.

Crowell *et al.* (2011) n'ont pas obtenu de bons ajustements aux données données indépendantes qui devaient servir à valider leur modèle. Pour expliquer cette différence, ils ont argumenté que les données provenant de l'étude de Wiersma et Roth (1983) auraient pu inclure des métabolites lipophiles en plus du BaP non métabolisé, des mesures au-dessous des limites de détection, la contamination d'échantillons, *etc.* De notre point de vue, il semble plus plausible que la différence de vitesse de métabolisation dans des conditions *in vivo* et *in vitro* soit la cause de ces disparités. Les valeurs de V_{max}/K_M utilisées dans le modèle de Crowell *et al.* (2011) proviennent de l'étude *in vitro* de Foth *et al.* (1988); elles n'apparaissent donc pas adéquates pour décrire correctement la biotransformation *in vivo* du BaP. Pery *et al.* (2011) en sont arrivés à la même conclusion lorsqu'ils ont été forcés de changer les paramètres de biotransformation hépatiques obtenus à partir des études *in vitro* pour simuler adéquatement les profils sanguins *in vivo* de Schlede *et al.* (1970).

Le taux d'élimination fécale ($0{,}27\ h^{-1}$) utilisé par Crowell *et al.* (2011) correspond au taux estimé par des relations empiriques basées sur des caractéristiques anatomiques et des débits physiologiques mesurés (Roth *et al.*, 1993). Le taux fécal estimé par notre modélisation a une valeur similaire : $0{,}33\ h^{-1}$.

Quant aux paramètres cinétiques du 3-OHBaP, l'absence de valeurs dans les publications scientifiques actuelles nous empêche de comparer les valeurs obtenues par notre modélisation cinétique.

7.2 Les résultats obtenus des modèles cinétiques

7.2.1 Cinétique du BaP

Les modèles que nous avons présentés dans ce manuscrit ont non seulement été utiles pour évaluer l'exposition des travailleurs dans leur milieu de travail, mais ils nous ont également permis de mieux comprendre la cinétique des deux molécules à l'étude.

L'adéquation entre les simulations du modèle et les profils sanguins et tissulaires observés dans les publications sélectionnées (voir tableau 5-2) vient corroborer la distribution rapide du BaP dans l'organisme tel qu'observée par Marie *et al.* (2010), Moir *et al.* (1998), Ramesh *et al.* (2001a) et Ramesh *et al.* (2001b). De plus, la modélisation PCBP montre qu'une métabolisation principalement par le foie peut expliquer les profils tissulaires et sanguins observés chez le rat. La modélisation à plusieurs compartiments montre aussi que les poumons peuvent également être considérés comme un site de biotransformation du BaP à un niveau moindre que le foie, tel que décrit par Wiersma et Roth (1983).

Grâce aux modèles cinétiques présentés et aux analyses de sensibilité réalisées, il a été constaté que la cinétique du BaP régit, en général, la cinétique du métabolite 3-OHBaP comme il a été observé pour le pyrène et le 1-OHP (Bouchard et Viau 1998). Également, il a été observé que la plupart des organes examinés sont en équilibre dynamique avec le sang et, donc, que les profils temporels du BaP et du 3-OHBaP dans plusieurs tissus évoluent en parallèle avec le profil sanguin peu après la période d'absorption comme observé par Crowell *et al.* (2011) et Moir *et al.* (1998) et par Bouchard et Viau (1998) pour le pyrène et le 1-OHP. De plus, il semble plausible que les composants lipidiques dans les organes agissent comme des compartiments de stockage de BaP, plus particulièrement, les tissus adipeux et les poumons (Marie *et al.* 2010). Par ailleurs, c'est dans les poumons que nous

227

retrouvons les plus grandes quantités de BaP, lorsqu'il a été administré par voie intraveineuse, en accord avec les résultats obtenus par Crowell *et al.* (2011), Moir *et al.* (1998) et Pery *et al.* (2011). Des quantités considérables de BaP se retrouvent également dans le foie et la peau. Finalement, la cinétique générale du BaP présente un comportement bi-exponentiel avec une métabolisation et une distribution rapide dans les premières vingt-quatre heures, suivie par une étape plus lente d'élimination due au relâchement lent du BaP des sites de stockage. Ce résultat confirme le comportement exponentiel observé et simulé lors des modélisations cinétiques décrites au tableau 2-12.

7.2.2 Cinétique du 3-OHBaP

La modélisation toxicocinétique a permis de constater que le 3-OHBaP est excrété en faible proportion dans l'urine et qu'un métabolisme essentiellement hépatique du BaP en 3-OHBaP peut être considéré pour simuler les données disponibles. D'un autre côté, les simulations cinétiques des profils temporels du 3-OHBaP suggèrent la présence d'un cycle entéro-hépatique (Chipman *et al.*, 1981b; Chipman *et al.*, 1981c; SCF, 2002). En l'absence de ce phénomène de réabsorption au niveau du tractus gastro-intestinal, les concentrations sanguines simulées pourraient difficilement atteindre les quantités observées dans les études sur les rats que nous avons examinées. D'ailleurs, c'est justement dans le foie que l'on retrouve des quantités élevées de 3-OHBaP (Cao *et al.*, 2005). Ceci corrobore le fait que le métabolite est produit dans le foie. De plus, le phénomène de recirculation du 3-OHBaP provenant du tractus gastro-intestinal peut faire augmenter les quantités de 3-OHBaP observées dans le foie.

Finalement, c'est au niveau des reins que les plus grandes quantités tissulaires de 3-OHBaP sont atteintes. Au cours de la période évaluée, les concentrations rénales de 3-

OHBaP étaient plus élevées que les concentrations hépatiques. Le contenu rénale de 3-OHBaP en fonction du temps ne semble pas être en équilibre dynamique avec l'évolution du contenu en 3-OHBaP dans le sang (Bouchard et Viau, 1997; Marie *et al.*, 2010). L'accumulation du 3-OHBaP explicable par une différence entre la vitesse d'entrée et de sortie des reins peut expliquer que les profils temporels sanguins et rénaux diffèrent (Bouchard et Viau, 1997).

7.3 Utilisation des modèles cinétiques dans la surveillance biologique du BaP

7.3.1 Reconstruction de la dose d'exposition

La surveillance biologique chez des groupes d'individus exposés au BaP est nécessaire pour estimer le risque des travailleurs exposés à des HAP cancérogènes. Or, d'une part, il est possible de demander aux travailleurs de porter des instruments de mesure des concentrations de BaP dans l'air avec lesquelles ils peuvent être en contact soit par inhalation (Ashok *et al.*, 2014; Chen *et al.*, 2012; Chen *et al.*, 2008; Devi *et al.*, 2013; Yamamoto *et al.*, 2014). D'autre part, il est possible de demander aux travailleurs de fournir des échantillons urinaires pour analyser le 3-OHBaP et, ensuite, calculer les concentrations de BaP correspondantes à l'exposition de chaque individu à travers la modélisation cinétique que nous avons décrite auparavant. Le présent travail permet d'utiliser cette dernière voie d'évaluation de l'exposition au BaP chez les travailleurs par une modélisation cinétique. Avec un outil de modélisation capable de représenter la cinétique du BaP et celle de son métabolite 3-OHBaP, il est possible de reconstruire les doses d'exposition au BaP correspondant aux profils d'excrétion urinaire du 3-OHBaP. Cependant, il est impératif que ces simulations se fassent à partir de collectes urinaires complètes (sans perte). Autrement, à partir de mesures ponctuelles, les profils d'exposition obtenus risquent de produire des

229

valeurs inadéquates étant donné la quantité d'information perdue dans les urines non-collectées.

Une approche probabiliste a été employée pour calculer les concentrations d'exposition au BaP (par voie respiratoire et par voie cutanée) puisque la résolution analytique de toutes les équations reliées au modèle PCBP était impraticable. Pour la détermination de valeurs de doses d'exposition de chaque individu, des techniques de simulation Monte-Carlo ont été utilisées. En général, les méthodes Monte-Carlo utilisées prenaient une série de valeurs au hasard d'un ensemble de paramètres choisis (par exemple, les différentes doses d'exposition pour chaque jour de travail). Ensuite, avec ces paramètres, les profils temporels de 3-OHBaP dans l'urine simulés à l'aide du modèle ont été comparés aux données observées pour chaque travailleur. Ce processus était exécuté pendant plusieurs cycles (de l'ordre de dix mille cycles par calcul) jusqu'à ce que le test statistique χ^2 de Pearson, calculé à partir des simulations, soit minimisé. Les concentrations de BaP obtenues par les simulations des profils d'excrétion urinaire de travailleurs étaient, en général, représentatives des scénarios d'exposition au BaP mesurés indépendamment. Ces scénarios d'exposition au BaP mesurés ont été établis à partir des formulaires remplis par les travailleurs concernant le temps de travail nécessaire pour réaliser leurs occupations professionnelles et par les concentrations atmosphériques mesurées de BaP.

Tel que présenté aux figures 4-8, 6-3, 6-4, et 6-5, les modèles cinétiques permettent d'estimer les concentrations de BaP nécessaires pour reproduire les profils d'excrétion urinaire collectés. Les valeurs de concentrations d'exposition au BaP simulées étaient du même ordre de grandeur que celles mesurées dans l'air, dans certains cas. Néanmoins, il y a eu d'autres cas où les concentrations obtenues avec les simulations n'ont pas pu être expliquées par l'absorption pulmonaire. Par exemple, l'excrétion urinaire continuait à augmenter lorsque les activités de travail étaient terminées. Un tel comportement cinétique pourrait être expliqué par une exposition cutanée plutôt que par l'inhalation du BaP. Tel que constaté par Forster *et al.* (2008), la voie principale d'exposition au BaP n'était pas l'inhalation lorsqu'une faible corrélation entre le BaP dans l'air ambiant et le 3-OHBaP dans l'urine était observée.

7.3.2 Comparaison de modèles toxicocinétiques

Les modèles à compartiments qui ont traditionnellement été utilisés dans le passé (Dvorchik et Vesell, 1976) ont été formés de un ou deux compartiments. Tel est le cas de modèles utilisés pour interpréter des données pour l'hydrazine, les dioxines (Lorber et Phillips, 2002) et le méthylmercure (Stern, 1997). Ainsi, lorsqu'on cherche à établir la vitesse d'élimination d'une substance toxique, un ensemble de fonctions physiologiques différentes peut être modélisé par un ou deux compartiments (Wells, 2012). L'avantage de cette méthode est la simplicité du modèle, ce qui conduit à des équations cinétiques simples avec des solutions pouvant être résolues analytiquement. Ce type de modèle possède l'avantage d'être construit, par définition, sur la base des données expérimentales obtenues sur des organismes *in vivo*. Donc, les incertitudes associées à l'extrapolation du comportement *in vitro* (Kedderis, 1997; Kedderis et Lipscomb, 2001; MacGregor *et al.*, 2001) sont réduites par rapport aux modèles PCBP (tel a été le cas du méthylmercure en 2001, NRC (2006)).

Dans le cas du BaP, plusieurs études, qui pourraient être classées dans la catégorie de modèles à compartiments, ont été réalisées. Néanmoins, dans la plupart de ces études (voir tableau 2-12), les chercheurs ont ajusté le profil sanguin à la somme de plusieurs fonctions exponentielles du type décrit par l'équation 3-1. Bien que cette technique permette de connaitre le taux d'élimination associé à la demi-vie du BaP, elle ne permet pas d'inférer la relation entre le profil sanguin du BaP et la dose d'exposition. Notre modélisation à un compartiment unique n'a pas seulement permis de déterminer les paramètres d'élimination, elle a aussi établi la fonction mathématique qui relie un profil d'exposition au BaP (plusieurs périodes d'exposition de différentes durées) au profil d'excrétion urinaire correspondant pour toutes les voies d'exposition.

Le modèle PCBP que nous avons décrit a également relié les profils d'exposition au BaP aux excrétions urinaires du 3-OHBaP chez les travailleurs. Cette modélisation physiologique a aussi permis de connaître les concentrations du BaP et du 3-OHBaP dans tous les organes simulés des travailleurs. Ceci est un avantage majeur de la modélisation PCBP par rapport à la modélisation à compartiments. Ainsi, si la connaissance de la cinétique est nécessaire pour déterminer l'organe cible du BaP ou les différences entre les concentrations tissulaires suite à des expositions par diverses voies, les modèles PCBP sont les meilleurs outils pour simuler le devenir du BaP ou de ses métabolites dans le corps. Cependant, en pratique, il est parfois difficile d'estimer les paramètres du modèle, puisqu'ils doivent être obtenus à partir de plusieurs études cinétiques (*in vivo* ou *in vitro* : Ekins, 2007) du composé toxique et de ses métabolites. En raison de la sensibilité de plusieurs paramètres, si on augmente le nombre de paramètres dans un modèle, on diminue sa fiabilité puisque ces modèles dépendent principalement des conditions dans lesquelles ils ont été construits (données expérimentales, nombre et choix des compartiments, détermination des coefficients de partition, des taux d'absorption, des taux de biotransformation, *etc.*). Telle a peut-être été la raison principale pour laquelle les simulations présentées et réalisées auparavant par les modèles PCBP n'ont pas réussi à reproduire les profils tissulaires des études indépendantes sans ajustement *a posteriori* (Crowell *et al.*, 2011 ; Pery *et al.*, 2011).

En comparant les deux modèles cinétiques développés dans le présent ouvrage, il a été constaté que les deux types de modélisation possèdent la capacité de simuler les profils d'excrétion urinaire de travailleurs, à partir de l'extrapolation des paramètres cinétiques de rats. Dans les deux cas, la reconstruction de doses d'exposition au BaP a été possible pour toutes les voies d'exposition et pour tous les travailleurs considérés dans trois sites industriels différents. Les principales différences entre les deux types de simulation résident dans la manière de construire chaque modèle et dans leur structure finale. En effet, les modèles reposent sur des philosophies différentes. D'une part, il est possible de définir les interactions individuelles de la substance toxique et de ses métabolites avec l'organisme, au niveau moléculaire (par exemple, le transport passif du BaP dans les tissus pulmonaires ou la perméabilité de la peau au BaP). D'autre part, le résultat global de ces processus individuels peuvent être observés à l'échelle de l'organisme au complet (comme les concentrations fécales du 3-OHBaP ou le taux d'absorption par la peau du BaP). De cette manière, les modèles à base physiologique tentent d'approcher la simulation mathématique avec un modèle idéalement générique, pour toute substance interagissant avec des processus physiologiques et mécanistes connus dans l'organisme, afin d'expliquer des mesures observées au niveau de matrices accessibles. Cette approche est de type « bottom-up ». En revanche, les modèles toxicocinétiques à compartiments sont spécifiques à une substance en particulier. Cette sorte de modélisation tente d'utiliser des mesures dans les matrices accessibles pour en reproduire d'autres à la même échelle et, par la suite, mieux comprendre les processus biologiques et les mécanismes impliqués dans les observations simulées. C'est ce qu'on appelle une approche de modélisation du type « top-down » (Wells, 2012).

Les deux types de modélisations se distinguent non seulement dans leur approche mais aussi dans le nombre de compartiments classiquement utilisés pour construire les modèles. Ainsi, les modèles PCBP sont généralement composés d'au moins quatre ou cinq compartiments alors que le modèle à compartiments simple peut être composé de seulement un compartiement. Par conséquent, les deux modèles utilisent un nombre très différent de paramètres. Les modèles PCBP les plus simples doivent au moins avoir une dizaine de paramètres tandis que les modèles à compartiments ont typiquement besoin de moins de dix

233

paramètres pour être fonctionnels. De ce fait, les simulations provenant de la modélisation PCBP peuvent consommer beaucoup de temps, comparativement aux simulations des modèles à compartiments. En effet, ces dernières nécessitent très peu de paramètres et utilisent souvent des équations résolues à l'avance. La simplicité des modèles à compartiments permet de résoudre le système d'équations différentielles propre à sa structure tandis que les équations provenant des modèles PCBP doivent être résolues numériquement pour chaque calcul réalisé par la simulation.

La nomenclature utilisée dans la modélisation PCBP est très claire puisqu'elle est standardisée tandis que la nomenclature de la modélisation à compartiments est très spécifique et souvent *ad hoc*. Par ailleurs, l'extrapolation inter-espèces est très simple et intuitivement logique pour les modèles PCBP tandis que les modèles à compartiments ont besoin de techniques heuristiques pour réaliser un changement entre espèces ou individus. Cependant, les deux types de modèles sont aussi aptes à réaliser des simulations déterministes ou stochastiques. Par simulation déterministe, on entend la simulation qui produit toujours le même résultat pour les mêmes données en traversant toujours la même séquence sous-jacente de calculs, tandis que la simulation stochastique est celle qui produit le résultat dans une plage de valeurs limitée en fonction d'une distribution de probabilité.

D'après les résultats obtenus, les deux modèles cinétiques sont appropriés pour reconstruire les doses d'exposition pour les voies principales d'absorption du BaP. À partir de ces résultats, il est acceptable de supposer que cette situation s'applique à d'autres produits toxiques. Cependant, puisque les modèles à base physiologique peuvent être très complexes dans leur structure, leur utilisation devrait se limiter aux cas où la reconstruction de la dose d'une composante nécessite une connaissance plus approfondie de la cinétique du composant ou de ses métabolites. Autrement, les modèles cinétiques à compartiments sont amplement suffisants pour décrire et relier les mesures au niveau des matrices accessibles sans s'occuper des processus mécanistes de façon superflue. Par exemple, dans le cas du BaP, notre modélisation n'a pas tenu compte du cœur de manière individuelle. Dans les deux modèles cinétiques, le cœur a été considéré comme faisant partie du « reste du corps » bien qu'il soit possible que les quantités de BaP retrouvées au cœur puissent être du même ordre de grandeur que dans les autres organes analysés. Ceci n'a pas empêché les modèles cinétiques utilisés de bien estimer les doses d'exposition des travailleurs, car la modélisation spécifique de la cinétique n'est pas importante pour déterminer la vitesse d'élimination du 3-OHBaP de l'organisme.

Les modèles cinétiques présentés ont toutefois des limites. Vu qu'une partie importante de la modélisation que nous avons réalisée vise à décrire la cinétique sur des sujets *in vivo*, nos modèles sont complètement basés sur des expériences réalisées *in vivo*. Lorsque les profils temporels du composé et ses métabolites d'intérêt dans le sang, les tissus et les excrétas d'animaux exposés ne sont pas disponibles, il est donc pratiquement impossible d'utiliser notre approche de modélisation par ajustement aux profils tissulaires ou d'excrétion. Également, les modèles PCBP et à compartiments proposés dépendent du type de données expérimentales disponibles dans les publications scientifiques. Ces études doivent présenter l'information nécessaire sur les profils tissulaires, les profils des excrétas accessibles, les voies d'exposition pertinentes, les périodes de temps adéquates et une puissance statistique des mesures expérimentales, en accord avec la précision du modèle souhaitée.

7.4 Conclusion

La modélisation cinétique proposée, en utilisant une approche de simulation basée sur des expériences sur des animaux *in vivo*, a permis de mieux comprendre la correspondance entre le BaP et son biomarqueur d'exposition, le 3-OHBaP. Comme il a été proposé pour d'autres molécules par notre équipe, la présente étude facilitera la détermination d'une valeur de référence biologique (VRB) pour le 3-OHBaP. Ainsi, il sera possible d'estimer la concentration maximale de 3-OHBaP dans l'urine correspondant aux valeurs de référence du BaP déterminées par les institutions règlementaires. Par exemple, l'excès de risque unitaire du BaP peut se traduire en quantité de BaP maximale par jour pour un adulte. Ainsi, avec une exposition chronique de BaP correspondant à cette quantité maximale, les modèles cinétiques peuvent calculer la concentration urinaire maximale de 3-OHBaP, au cours de la durée de l'exposition.

Par ailleurs, les principales applications de modèles cinétiques ne se restreignent pas uniquement à la reconstruction de doses absorbées de xénobiotiques par des mesures biologiques répétées, mais également à l'estimation des voies principales d'entrée en fonction des tâches de travail. Dans le cas du BaP, cette dernière application a été plus difficile à réaliser puisque nous avons utilisé seulement un biomarqueur pour déterminer les scenarios d'exposition au BaP (les doses et les durées d'exposition). Or, avec l'utilisation de deux ou plusieurs métabolites pour estimer les profils d'exposition au BaP, la détermination de la voie d'exposition serait sans équivoque et simple, d'où l'importance d'utiliser au moins deux biomarqueurs pour prédire des scénarios d'exposition réalistes dans le milieu industriel.

Enfin, la comparaison de l'utilisation de modèles cinétiques à base physiologique et de modèles à compartiments en surveillance biologique nous a permis de valider une nouvelle alternative aux modèles PCBP traditionnellement utilisés. Ainsi, notre étude sera très utile à beaucoup d'autres expériences futures sur d'autres substances et permettra aux chercheurs d'avoir une alternative beaucoup plus simple et fiable grâce à l'utilisation de modèles simples à compartiments. Évidemment, le choix final entre l'utilisation de modèles

PCBP ou à compartiments dépend des besoins particuliers concernant la substance toxique à l'étude.

8 Bibliographie

Aarstad K, Toftgard R and Nilsen OG. (1987) A comparison of the binding and distribution of benzo[a]pyrene in human and rat serum. *Toxicology* 47: 235-245.

Agbato I. (2006) *Établissement de la cinétique d'excrétion urinaire du 3-hydroxybenzo(a)pyrène suite à l'injection de faibles doses de benzo(a)pyrène chez le rat Sprague-Dawley mâle.*

Aiache JM. (1985) *Traité de biopharmacie et pharmacocinétique,* Paris Montréal: Vigot ; Presses de l'Université de Montréal.

Albanese RA, Banks HT, Evans MV, *et al.* (2002) Physiologically based pharmacokinetic models for the transport of trichloroethylene in adipose tissue. *Bulletin of Mathematical Biology* 64: 97-131.

Allamandola LJ. (2011) Pahs and Astrobiology. *Pahs and the Universe: A Symposium to Celebrate the 25th Anniversary of the Pah Hypothesis* 46: 305-317.

Anakwe K, Robson L, Hadley J, *et al.* (2002) 16 Wnt regulation of limb muscle differentiation. *Journal of anatomy* 201: 421.

Andersen ME. (1995) Development of Physiologically-Based Pharmacokinetic and Physiologically-Based Pharmacodynamic Models for Applications in Toxicology and Risk Assessment. *Toxicology Letters* 79: 35-44.

Andersen ME, Clewell HJ, Gargas ML, *et al.* (1987) Physiologically Based Pharmacokinetics and the Risk Assessment Process for Methylene-Chloride. *Toxicology and Applied Pharmacology* 87: 185-205.

Anderson LM, Ruskie S, Carter J, *et al.* (1995) Fetal mouse susceptibility to transplacental carcinogenesis: Differential influence of Ah receptor phenotype on effects of 3-methylcholanthrene, 12-dimethylbenz[a]anthracene, and benzo[a]pyrene. *Pharmacogenetics* 5: 364-372.

Angerer J and Deutsche Forschungsgemeinschaft. (2006) *Essential biomonitoring methods from the MAK-Collection for occupational health and safety,* Weinheim: Wiley-VCH.

Arfken GB, Weber H-J, Harris FE, *et al.* (2012) Mathematical methods for physicists. 7th ed. Oxford: Academic,, 1 online resource (1 v.).

Ariese F, Verkaik M, Hoornweg GP, *et al.* (1994) Trace Analysis of 3-Hydroxy Benzo[a]Pyrene in Urine for the Biomonitoring of Human Exposure to Polycyclic Aromatic-Hydrocarbons. *Journal of Analytical Toxicology* 18: 195-204.

Armstrong BG and Gibbs G. (2009) Exposure-response relationship between lung cancer and polycyclic aromatic hydrocarbons (PAHs). *Occupational and Environmental Medicine* 66: 740-746.

Ashok V, Gupta T, Dubey S, *et al.* (2014) Personal exposure measurement of students to various microenvironments inside and outside the college campus. *Environmental Monitoring and Assessment* 186: 735-750.

Atkins GL. (1969) *Multicompartment models for biological systems,* London,: Methuen.

ATSDR. (1995a) Toxicological Profile for Polycyclic Aromatic Hydrocarbons. In: SERVICES USDOHAH (ed). Agency for Toxic Substances and Disease Registry.

ATSDR. (1995b) *Toxicological profile for polycyclic aromatic hydrocarbons,* Atlanta, GA: U.S. Department of Health and Human Services.

ATSDR. (2011) *The ATSDR 2011 Substance Priority List.* Available at: http://www.atsdr.cdc.gov/spl/.

Autrup H, Grafstrom RC, Brugh M, *et al.* (1982) Comparison of benzo(a)pyrene metabolism in bronchus, esophagus, colon, and duodenum from the same individual. *Cancer Res* 42: 934-938.

Awogi T and Sato T. (1989) Micronucleus test with benzo[a]pyrene using a single peroral administration and intraperitoneal injection in males of the MS/Ae and CD-1 mouse strains. *Mutat Res* 223: 353-356.

Baan R, Grosse Y, Straif K, *et al.* (2009) A review of human carcinogens--Part F: chemical agents and related occupations. *The lancet oncology* 10: 1143-1144.

Balant LP and Gexfabry M. (1990) Physiological Pharmacokinetic Modeling. *Xenobiotica; the fate of foreign compounds in biological systems* 20: 1241-1257.

Berg Hvd. (2011) *Mathematical models of biological systems,* Oxford: Oxford University Press.

Berthet A, Heredia-Ortiz R, Vernez D, *et al.* (2012) A detailed urinary excretion time course study of captan and folpet biomarkers in workers for the estimation of dose, main route-of-entry and most appropriate sampling and analysis strategies. *The Annals of occupational hygiene* 56: 815-828.

Blanton RH, Lyte M, Myers MJ, *et al.* (1986) Immunomodulation by polyaromatic hydrocarbons in mice and murine cells. *Cancer Res* 46: 2735-2739.

Bostrom CE, Gerde P, Hanberg A, *et al.* (2002) Cancer risk assessment, indicators, and guidelines for polycyclic aromatic hydrocarbons in the ambient air. *Environ Health Perspect* 110 Suppl 3: 451-488.

Bouchard M, Brunet RC, Droz P-O, Carrier G. A biologically-based dynamic model for predicting the disposition of methanol and its metabolites in animals and humans. Toxicological Sciences 64: 169-184, 2001.

Bouchard M, Carrier G, Brunet RC, *et al.* (2005) Determination of biological reference values for chlorpyrifos metabolites in human urine using a toxicokinetic approach. *Journal of Occupational and Environmental Hygiene* 2: 155-168.

Bouchard M, Carrier G, Brunet R., Dumas P, Noisel N. Biological monitoring of exposure to organophosphorus insecticides in horticultural greenhouse workers. Annals of Occupational Hygiene 50(5): 505-515, 2006.

Bouchard M, Dodd C and Viau C. (1994) Improved Procedure for the High-Performance Liquid-Chromatographic Determination of Monohydroxylated Pah Metabolites in Urine. *Journal of Analytical Toxicology* 18: 261-264.

Bouchard M and Viau C. (1997) Urinary excretion of benzo[a]pyrene metabolites following intravenous, oral, and cutaneous benzo[a]pyrene administration. *Can J Physiol Pharmacol* 75: 185-192.

Bouchard M, Viau C, Marie M, *et al.* (2010) Apport de la toxicocinétique dans le suivi biologique des expositions aux hydrocarbures aromatiques polycycliques. *Archives des Maladies Professionnelles* 71: 520-522.

Brauer F. (2008) Compartmental models in epidemiology. *Mathematical Epidemiology* 1945: 19-79.

Brinkmann J, Stolpmann K, Trappe S, *et al.* (2013) Metabolically Competent Human Skin Models: Activation and Genotoxicity of Benzo[a]pyrene. *Toxicological Sciences* 131: 351-359.

Brown RP, Delp MD, Lindstedt SL, *et al.* (1997) Physiological parameter values for physiologically based pharmacokinetic models. *Toxicology and Industrial Health* 13: 407-484.

Burstyn I, Kromhout H, Johansen C, *et al.* (2007) Bladder cancer incidence and exposure to polycyclic aromatic hydrocarbons among asphalt pavers. *Occupational and Environmental Medicine* 64: 520-526.

Campo L, Rossella F, Pavanello S, *et al.* (2010) Urinary profiles to assess polycyclic aromatic hydrocarbons exposure in coke-oven workers. *Toxicology Letters* 192: 72-78.

Cao D, Yoon CH, Shin BS, *et al.* (2005) Effects of aloe, aloesin, or propolis on the pharmacokinetics of benzo[a]pyrene and 3-OH-benzo[a]pyrene in rats. *Journal of toxicology and environmental health. Part A* 68: 2227-2238.

Carrell CJ, Carrell TG, Carrell HL, *et al.* (1997) Benzo[a]pyrene and its analogues: structural studies of molecular strain. *Carcinogenesis* 18: 415-422.

Carrier G, Bouchard M, Brunet RC, Caza M. A toxicokinetic model for predicting the tissue distribution and elimination of organic and inorganic mercury following exposure to methyl mercury in animals and humans. II- Application and validation of the model in humans. Toxicology Applied Pharmacology 171: 50-60, 2001.

Casarett LJ, Klaassen CD, Amdur MO, *et al.* (1996) *Casarett and Doull's toxicology : the basic science of poisons,* New York: McGraw-Hill, Health Professions Division.

Chen C, Campbell KD, Negi I, *et al.* (2012) A New Sensor for the Assessment of Personal Exposure to Volatile Organic Compounds. *Atmos Environ (1994)* 54: 679-687.

Chen MR, Tsai PJ and Wang YF. (2008) Assessing inhalatory and dermal exposures and their resultant health-risks for workers exposed to polycyclic aromatic hydrocarbons (PAHs) contained in oil mists in a fastener manufacturing industry. *Environ Int* 34: 971-975.

Chien YC and Yeh CT. (2012) Excretion kinetics of urinary 3-hydroxybenzo[a]pyrene following dietary exposure to benzo[a]pyrene in humans. *Archives of Toxicology* 86: 45-53.

Chipman JK, Bhave NA, Hirom PC, *et al.* (1982) Metabolism and excretion of benzo[a]pyrene in the rabbit. *Xenobiotica; the fate of foreign compounds in biological systems* 12: 397-404.

Chipman JK, Frost GS, Hirom PC, *et al.* (1981a) Biliary excretion, systemic availability and reactivity of metabolites following intraportal infusion of [3H]benzo[a]pyrene in the rat. *Carcinogenesis* 2: 741-745.

Chipman JK, Hirom PC, Frost GS, *et al.* (1981b) Benzo(a)pyrene metabolism and enterohepatic circulation in the rat. *Adv Exp Med Biol* 136 Pt A: 761-768.

Chipman JK, Hirom PC, Frost GS, *et al.* (1981c) The biliary excretion and enterohepatic circulation of benzo(a)pyrene and its metabolites in the rat. *Biochem Pharmacol* 30: 937-944.

Ciffroy P, Tanaka T, Johansson E, *et al.* (2011) Linking fate model in freshwater and PBPK model to assess human internal dosimetry of B(a)P associated with drinking water. *Environmental geochemistry and health* 33: 371-387.

CIRC. (1983) *Polynuclear aromatic compounds, part 1 : chemical environmental and experimental data,* Lyon: International Agency for Research on Cancer.

Coelho E, Ferreira C and Almeida CMM. (2008) Analysis of polynuclear aromatic hydrocarbons by SPME-GC-FID in environmental and tap waters. *Journal of the Brazilian Chemical Society* 19: 1084-1097.

Conney AH, Miller EC and Miller JA. (1956) The metabolism of methylated aminoazo dyes. V. Evidence for induction of enzyme synthesis in the rat by 3-methylcholanthrene. *Cancer Res* 16: 450-459.

Cote J, Bonvalot Y, Carrier G, *et al.* (2014) A novel toxicokinetic modeling of cypermethrin and permethrin and their metabolites in humans for dose reconstruction from biomarker data. *PLoS One* 9: e88517.

Cottini GB and Mazzone GB. (1939) The efects of 3,4-Benzopyrene on human skin. *American Journal of Cancer* 37.

Crowell SR, Amin SG, Anderson KA, *et al.* (2011) Preliminary physiologically based pharmacokinetic models for benzo[a]pyrene and dibenzo[def,p]chrysene in rodents. *Toxicol Appl Pharmacol* 257: 365-376.

Culp SJ, Gaylor DW, Sheldon WG, *et al.* (1998) A comparison of the tumors induced by coal tar and benzo[a]pyrene in a 2-year bioassay. *Carcinogenesis* 19: 117-124.

Davies B and Morris T. (1993) Physiological-Parameters in Laboratory-Animals and Humans. *Pharmaceutical Research* 10: 1093-1095.

Dean RG, Bynum G, Jacobson-Kram D, *et al.* (1983) Activation of polycyclic hydrocarbons in Reuber H4-II-E hepatoma cells. An in vitro system for the induction of SCEs. *Mutat Res* 111: 419-427.

Devi JJ, Gupta T, Jat R, *et al.* (2013) Measurement of personal and integrated exposure to particulate matter and co-pollutant gases: a panel study. *Environ Sci Pollut Res Int* 20: 1632-1648.

deVries A, vanOostrom CTM, Dortant PM, *et al.* (1997) Spontaneous liver tumors and benzo[a]pyrene-induced lymphomas in XPA-deficient mice. *Molecular Carcinogenesis* 19: 46-53.

Druckrey H, Preussmann R, Ivankovic S, *et al.* (1967) [Organotropic carcinogenic effects of 65 various N-nitroso- compounds on BD rats]. *Z Krebsforsch* 69: 103-201.

Dvorchik BH and Vesell ES. (1976) Pharmacokinetic Interpretation of Data Gathered during Therapeutic Drug Monitoring. *Clinical Chemistry* 22: 868-878.

EC. (2002) Polycyclic Aromatic Hydrocarbons – Occurrence in foods, dietary exposure and health effects. In: Commission E (ed). European Commission.

Ekins S. (2007) *Computational toxicology : risk assessment for pharmaceutical and environmental chemicals,* Hoboken, N.J.: Wiley-Interscience.

EPA. (1984) Health Effects Assessment for Benzo[a]pyrene. In: U.S. Environmental Protection Agency (ed). Washington, D.C.: U.S. Environmental Protection Agency,
.

EPA. (1990) *Toxicological profile for benzo a pyrene,* Atlanta, Georgia: U.S. Department of Health & Human Services Public Health Service Agency for Toxic Substances and Disease Registry.

EPA. (2014) *ELECTRONIC CODE OF FEDERAL REGULATIONS.* Available at: http://water.epa.gov/scitech/methods/cwa/pollutants.cfm.

Espie P, Tytgat D, Sargentini-Maier ML, *et al.* (2009) Physiologically based pharmacokinetics (PBPK). *Drug Metabolism Reviews* 41: 391-407.

Forster K, Preuss R, Rossbach B, *et al.* (2008) 3-Hydroxybenzo[a]pyrene in the urine of workers with occupational exposure to polycyclic aromatic hydrocarbons in different industries. *Occupational and Environmental Medicine* 65: 224-229.

Foth H, Kahl R and Kahl GF. (1988) Pharmacokinetics of Low-Doses of Benzo[a]Pyrene in the Rat. *Food and Chemical Toxicology* 26: 45-51.

Fox CH, Selkirk JK, Price FM, *et al.* (1975) Metabolism of benzo(a)pyrene by human epithelial cells in vitro. *Cancer Res* 35: 3551-3557.

Garrett ER. (1994) Simplified Methods for the Evaluation of the Parameters of the Time-Course of Plasma-Concentration in the One-Compartment Body Model with First-Order Invasion and First-Order Drug Elimination Including Methods for Ascertaining When Such Rate Constants Are Equal (Vol 21, Pg 689, 1993). *Journal of Pharmacokinetics and Biopharmaceutics* 22: 179-179.

Gelboin HV. (1980) Benzo[a]Pyrene Metabolism, Activation, and Carcinogenesis - Role and Regulation of Mixed-Function Oxidases and Related Enzymes. *Physiological Reviews* 60: 1107-1166.

Gendre C, Lafontaine M, Delsaut P, *et al.* (2004) Exposure to polycyclic aromatic hydrocarbons and excretion of urinary 3-hydroxybenzo[A]pyrene: Assessment of an appropriate sampling time. *Polycyclic Aromatic Compounds* 24: 433-439.

Genium Publishing Corporation. (1999) *Genium's handbook of safety, health, and environmental data for common hazardous substances,* New York: McGraw Hill.

Gerde P and Scholander P. (1988) Adsorption of Benzo(a)Pyrene on to Asbestos and Manmade Mineral Fibers in an Aqueous-Solution and in a Biological Model Solution. *British Journal of Industrial Medicine* 45: 682-688.

Gerlowski LE and Jain RK. (1983) Physiologically based pharmacokinetic modeling: principles and applications. *Journal of pharmaceutical sciences* 72: 1103-1127.

Ginsberg GL, Atherholt TB and Butler GH. (1989) Benzo[a]pyrene-induced immunotoxicity: comparison to DNA adduct formation in vivo, in cultured splenocytes, and in microsomal systems. *J Toxicol Environ Health* 28: 205-220.

Godfrey K. (1983) *Compartmental models and their application,* London ; New York: Academic Press.

Goksoyr A. (1995) Use of Cytochrome-P450-1a (Cyp1a) in Fish as a Biomarker of Aquatic Pollution. *Toxicology in Transition*: 80-95.

Gosselin NH, Brunet RC, Carrier G, *et al.* (2006) Reconstruction of methylmercury intakes in indigenous populations from biomarker data. *Journal of exposure science & environmental epidemiology* 16: 19-29.

Gosselin NH, Brunet RC, Carrier G, *et al.* (2005) Worker exposures to triclopyr: risk assessment through measurements in urine samples. *The Annals of occupational hygiene* 49: 415-422.

Goulding R. (1986) Environmental-Health Criteria, Vol 57, Principles of Toxicokinetic Studies. *Journal of the Royal Society of Health* 106: 228-228.

Gundel J, Schaller KH and Angerer J. (2000) Occupational exposure to polycyclic aromatic hydrocarbons in a fireproof stone producing plant: biological monitoring of 1-hydroxypyrene, 1-, 2-, 3- and 4-hydroxyphenanthrene, 3-hydroxybenz(a)anthracene and 3-hydroxybenzo(a)pyrene. *International archives of occupational and environmental health* 73: 270-274.

Gupta P, Banerjee DK, Bhargava SK, *et al.* (1993) Prevalence of Impaired Lung Function in Rubber Manufacturing Factory Workers Exposed to Benzo(a)pyrene and Respirable Particulate Matter. *Indoor and Built Environment* 2: 26-31.

Haines PA and Hendrickson MD. (2009) *Polycyclic aromatic hydrocarbons : pollution, health effects and chemistry,* New York: Nova Science Publishers.

Hammond EC, Selikoff IJ, Lawther PL, *et al.* (1976) Inhalation of Benzpyrene and Cancer in Man. *Annals of the New York Academy of Sciences* 271: 116-124.

Harrigan JA, McGarrigle BP, Sutter TR, *et al.* (2006) Tissue specific induction of cytochrome P450 (CYP) 1A1 and 1B1 in rat liver and lung following in vitro (tissue slice) and in vivo exposure to benzo(a)pyrene. *Toxicology in Vitro* 20: 426-438.

Harris CC, Autrup H, Stoner GD, *et al.* (1979) Metabolism of benzo(a)pyrene, N-nitrosodimethylamine, and N-nitrosopyrrolidine and identification of the major carcinogen-DNA adducts formed in cultured human esophagus. *Cancer Res* 39: 4401-4406.

243

Harvey RG. (1991) *Polycyclic aromatic hydrocarbons : chemistry and carcinogenicity,* Cambridge ; New York: Cambridge University Press.

Harvey RG. (1997) *Polycyclic aromatic hydrocarbons,* New York: Wiley-VCH.

Hauck M, Huijbregts MAJ, Armitage JM, *et al.* (2008) Model and input uncertainty in multimedia fate modeling: Benzo[a]pyrene concentrations in Europe. *Chemosphere* 72: 959-967.

Hawkins WE, Walker WW, Overstreet RM, *et al.* (1988) Dose-Related Carcinogenic Effects of Water-Borne Benzo[a]Pyrene on Livers of 2 Small Fish Species. *Ecotoxicology and Environmental Safety* 16: 219-231.

Hayes AW. (2008) *Principles and methods of toxicology,* Boca Raton: CRC Press.

He SX, Nicholson RA and Law FCP. (1998) Benzo(a)pyrene toxicokinetics in the cricket following injection into the haemolymph. *Environmental Toxicology and Pharmacology* 6: 81-89.

Health Canada. (1994) *Polycyclic aromatic hydrocarbons,* Ottawa: Govt. of Canada Health Canada.

Heinonen J, Kukkonen JVK and Holopainen IJ. (2000) Toxiceokinetics of 2,4,5-trichlorophenol and benzo(a)pyrene in the clam Pisidium amnicum: Effects of seasonal temperatures and trematode parasites. *Archives of Environmental Contamination and Toxicology* 39: 352-359.

Hendricks JD, Meyers TR, Shelton DW, *et al.* (1985) Hepatocarcinogenicity of Benzo[a]Pyrene to Rainbow-Trout by Dietary Exposure and Intraperitoneal Injection. *Journal of the National Cancer Institute* 74: 839-851.

Heredia-Ortiz R, Berthet A and Bouchard M. (2011) Toxicokinetic modeling of folpet fungicide and its ring-biomarkers of exposure in humans. *Journal of applied toxicology : JAT.*

Heredia-Ortiz R and Bouchard M. (2012) Toxicokinetic modeling of captan fungicide and its tetrahydrophthalimide biomarker of exposure in humans. *Toxicology Letters* 213: 27-34.

Hudgins DM. (2002) Interstellar polycyclic aromatic compounds and astrophysics. *Polycyclic Aromatic Compounds* 22: 469-488.

IARC. (1983) Polynuclear aromatic compound: Part 1, Chemical, Environmental and Experimental Data. In: humans IAfRoCmoteocrt (ed). Lyon, France: International Agency for Research on Cancer, 1-457.

IARC. (2010) *Some non-heterocyclic polycyclic aromatic hydrocarbons and some related occupational exposures,* Lyon, France, Geneva: IARC Press. Distributed by World Health Organization.

IPCS. (2010) Characterization and application of physiologically based pharmacokinetic models in risk assessment. In: Safety. WHOIPoC (ed). Harmonization project document no. 9.

IRIS. (2014) *IRIS Toxicological Review of Benzo[a]pyrene (Public Comment Draft) posted.* Available at: http://www.epa.gov/IRIS/.

Jacques C, Perdu E, Duplan H, *et al.* (2010) Disposition and biotransformation of C-14-Benzo(a)pyrene in a pig ear skin model: Ex vivo and in vitro approaches. *Toxicology Letters* 199: 22-33.

Jacquez JA. (1985) *Compartmental analysis in biology and medicine,* Ann Arbor: University of Michigan Press.

James MO, Altman AH, Li CLJ, *et al.* (1995) Biotransformation, Hepatopancreas DNA-Binding and Pharmacokinetics of Benzo[a]Pyrene after Oral and Parenteral Administration to the American Lobster, Homarus-Americanus. *Chem Biol Interact* 95: 141-160.

James MO, Tong Z, Rowland-Faux L, *et al.* (2001) Intestinal bioavailability and biotransformation of 3-hydroxybenzo(a)pyrene in an isolated perfused preparation from channel catfish, Ictalurus punctatus. *Drug metabolism and disposition: the biological fate of chemicals* 29: 721-728.

Joblin C and Mulas G. (2009) Interstellar Polycylic Aromatic Hydrocarbons: From Space to the Laboratory. *Interstellar Dust from Astronomical Observations to Fundamental Studies* 35: 133-152.

Jongeneelen FJ, Anzion RBM and Henderson PT. (1987) Determination of Hydroxylated Metabolites of Polycyclic Aromatic-Hydrocarbons in Urine. *Journal of Chromatography-Biomedical Applications* 413: 227-232.

Jongeneelen FJ, Bos RP, Anzion RB, *et al.* (1986) Biological monitoring of polycyclic aromatic hydrocarbons. Metabolites in urine. *Scandinavian journal of work, environment & health* 12: 137-143.

Jongeneelen FJ, Leijdekkers CM, Bos RP, *et al.* (1985) Excretion of 3-hydroxy-benzo(a)pyrene and mutagenicity in rat urine after exposure to benzo(a)pyrene. *Journal of applied toxicology : JAT* 5: 277-282.

Jongeneelen FJ, Leijdekkers CM and Henderson PT. (1984) Urinary excretion of 3-hydroxy-benzo[a]pyrene after percutaneous penetration and oral absorption of benzo[a]pyrene in rats. *Cancer Lett* 25: 195-201.

Kao J, Patterson FK and Hall J. (1985) Skin penetration and metabolism of topically applied chemicals in six mammalian species, including man: an in vitro study with benzo[a]pyrene and testosterone. *Toxicol Appl Pharmacol* 81: 502-516.

Karlehagen S, Andersen A and Ohlson CG. (1992) Cancer Incidence among Creosote-Exposed Workers. *Scandinavian Journal of Work Environment & Health* 18: 26-29.

Kawamura Y, Kamata E, Ogawa Y, *et al.* (1988) The Effect of Various Foods on the Intestinal-Absorption of Benzo(a)Pyrene in Rats. *Journal of the Food Hygienic Society of Japan* 29: 21-25.

Kedderis GL. (1997) Extrapolation of in vitro enzyme induction data to humans in vivo. *Chem Biol Interact* 107: 109-121.

Kedderis GL and Lipscomb JC. (2001) Application of in vitro biotransformation data and pharmacokinetic modeling to risk assessment. *Toxicology and Industrial Health* 17: 315-321.

245

Kim JH, Stansbury KH, Walker NJ, *et al.* (1998) Metabolism of benzo[a]pyrene and benzo[a]pyrene-7,8-diol by human cytochrome P450 1B1. *Carcinogenesis* 19: 1847-1853.

Klaassen CD. (2013) *Casarett and Doull's toxicology : the basic science of poisons,* New York: McGraw-Hill, Medical Publishing Division.

Knaak JB, Timchalk C, Tornero-Velez R, et al. (2012) Parameters for pesticide QSAR and PBPK/PD models for human risk assessment, Washington, DC: American Chemical Society.

Knudsen L, Merlo DF and Royal Society of Chemistry (Great Britain). (2011a) Biomarkers and human biomonitoring. Vol. 1, Ongoing programs and exposures. *Issues in toxicology.* Cambridge: Royal Society of Chemistry,, 1 online resource (1 v.).

Knudsen L, Merlo DF and Royal Society of Chemistry (Great Britain). (2011b) Biomarkers and human biomonitoring. Vol. 2, Selected biomarkers of current interest. *Issues in toxicology.* Cambridge: Royal Society of Chemistry,, 1 online resource (1 v.).

Krishnan K and Andersen ME. (2010) *Quantitative modeling in toxicology,* Chichester, West Sussex Hoboken: John Wiley & Sons.

Krishnan K, Clewell HJ, 3rd and Andersen ME. (1994) Physiologically based pharmacokinetic analyses of simple mixtures. *Environ Health Perspect* 102 Suppl 9: 151-155.

Lafontaine M, Gendre C, Delsaut P, *et al.* (2004) Urinary 3-hydroxybenzo[A]pyrene as a biomarker of exposure to polycyclic aromatic hydrocarbons: An approach for determining a biological limit value. *Polycyclic Aromatic Compounds* 24: 441-450.

Lafontaine M, Payan JP, Delsaut P, *et al.* (2000) Polycyclic aromatic hydrocarbon exposure in an artificial shooting target factory: assessment of 1-hydroxypyrene urinary excretion as a biological indicator of exposure. *The Annals of occupational hygiene* 44: 89-100.

Lamka J, Rudisar L and Kvetina J. (1991) On the limiting factors affecting the distribution of model drugs from blood into the lymphatic system. *Eur J Drug Metab Pharmacokinet* Spec No 3: 47-51.

Landrum PF. (1989) Bioavailability and Toxicokinetics of Polycyclic Aromatic-Hydrocarbons Sorbed to Sediments for the Amphipod Pontoporeia-Hoyi. *Environmental Science & Technology* 23: 588-595.

Laskarzewski PM, Weiner DL and Ott L. (1982) A Simulation Study of Parameter-Estimation in the One and 2 Compartment Models. *Journal of Pharmacokinetics and Biopharmaceutics* 10: 317-334.

Laurent C, Feidt C, Lichtfouse E, *et al.* (2001) Milk-blood transfer of C-14-Tagged polycyclic aromatic hydrocarbons (PAHs) in pigs. *Journal of Agricultural and Food Chemistry* 49: 2493-2496.

Lavoie EJ, Braley J, Rice JE, *et al.* (1987) Tumorigenic activity of non-alternant polynuclear aromatic hydrocarbons in newborn mice. *Cancer Lett* 34: 15-20.

Lee W, Shin HS, Hong JE, *et al.* (2003) Studies on the analysis of benzo(a)pyrene and its metabolites in biological samples by using high performance liquid chromatograpy/fluorescence detection and gas chromatography/mass spectrometry. *Bulletin of the Korean Chemical Society* 24: 559-565.

Leeuwen CJv and Vermeire T. (2007) *Risk assessment of chemicals : an introduction,* Dordrecht: Springer.

Legraverend C, Guenthner TM and Nebert DW. (1984) Importance of the route of administration for genetic differences in benzo[a]pyrene-induced in utero toxicity and teratogenicity. *Teratology* 29: 35-47.

Leonov GA. (2001) *Mathematical problems of control theory : an introduction,* Singapore ; River Edge, N.J.: World Scientific.

Leroyer A, Jeandel F, Maitre A, *et al.* (2010) 1-Hydroxypyrene and 3-hydroxybenzo[a]pyrene as biomarkers of exposure to PAH in various environmental exposure situations. *Science of the Total Environment* 408: 1166-1173.

Leversee GJ, Giesy JP, Landrum PF, *et al.* (1982) Kinetics and Biotransformation of Benzo(a)Pyrene in Chironomus-Riparius. *Archives of Environmental Contamination and Toxicology* 11: 25-31.

Likhachev AJ, Beniashvili D, Bykov VJ, *et al.* (1992) Biomarkers for individual susceptibility to carcinogenic agents: excretion and carcinogenic risk of benzo[a]pyrene metabolites. *Environ Health Perspect* 98: 211-214.

Lipscomb JC and Ohanian EV. (2007) *Toxicokinetics and risk assessment,* New York: Informa Healthcare.

Liu D, Pan L, Yang H, *et al.* (2014) A physiologically based toxicokinetic and toxicodynamic model links the tissue distribution of benzo[a]pyrene and toxic effects in the scallop Chlamys farreri. *Environ Toxicol Pharmacol* 37: 493-504.

Loeb LA and Harris CC. (2008) Advances in chemical carcinogenesis: a historical review and prospective. *Cancer Res* 68: 6863-6872.

Lorber M and Phillips L. (2002) Infant exposure to dioxin-like compounds in breast milk. *Environ Health Perspect* 110: A325-A332.

Lu XX, Reible DD and Fleeger JW. (2004) Bioavailability and assimilation of sediment-associated benzo[a]pyrene by Ilyodrilus templetoni (oligochaeta). *Environmental Toxicology and Chemistry* 23: 57-64.

Luch A. (2005) *The carcinogenic effects of polycyclic aromatic hydrocarbons,* London, Hackensack, NJ ; London: Imperial College Press.

MacGregor JT, Collins JM, Sugiyama Y, *et al.* (2001) In vitro human tissue models in risk assessment: report of a consensus-building workshop. *Toxicological Sciences* 59: 17-36.

Maclure KM and MacMahon B. (1980) An epidemiologic perspective of environmental carcinogenesis. *Epidemiol Rev* 2: 19-48.

Madhavan ND and Naidu KA. (2000) Purification and partial characterization of peroxidase from human term placenta of non-smokers: Metabolism of benzo(a)pyrene-7,8-dihydrodiol. *Placenta* 21: 501-509.

Marie C, Bouchard M, Heredia-Ortiz R, *et al.* (2010) A toxicokinetic study to elucidate 3-hydroxybenzo(a)pyrene atypical urinary excretion profile following intravenous injection of benzo(a)pyrene in rats. *Journal of applied toxicology : JAT* 30: 402-410.

Mazumdar S, Redmond C, Sollecito W, *et al.* (1975) Epidemiological Study of Exposure to Coal-Tar Pitch Volatiles among Coke Oven Workers. *Journal of the Air Pollution Control Association* 25: 382-389.

McCormick DL, Burns FJ and Albert RE. (1981) Inhibition of benz[a]pyrene-induced mammary carcinogenesis by retinyl acetate. *J Natl Cancer Inst* 66: 559-564.

McNally K, Cotton R, Cocker J, *et al.* (2012) Reconstruction of Exposure to m-Xylene from Human Biomonitoring Data Using PBPK Modelling, Bayesian Inference, and Markov Chain Monte Carlo Simulation. *Journal of toxicology* 2012: 760281.

McNally K, Cotton R and Loizou GD. (2011) A Workflow for Global Sensitivity Analysis of PBPK Models. *Frontiers in pharmacology* 2: 31.

Mehta R, Meredith-Brown M and Cohen GM. (1979) Metabolism and covalent binding of benzo[alpha]pyrene in human peripheral lung. *Chem Biol Interact* 28: 345-358.

Meibohm B and Derendorf H. (1997) Basic concepts of pharmacokinetic/pharmacodynamic (PK/PD) modelling. *International journal of clinical pharmacology and therapeutics* 35: 401-413.

Moir D, Viau A, Chu I, *et al.* (1998) Pharmacokinetics of benzo[a]pyrene in the rat. *Journal of toxicology and environmental health. Part A* 53: 507-530.

Molokanov A, Chojnacki E and Blanchardon E. (2010) A simple algorithm for solving the inverse problem of interpretation of uncertain individual measurements in internal dosimetry. *Health Phys* 98: 12-19.

Moody RP, Joncas J, Richardson M, *et al.* (2007) Contaminated soils (I): In vitro dermal absorption of benzo[a] pyrene in human skin. *Journal of Toxicology and Environmental Health-Part a-Current Issues* 70: 1858-1865.

Moore BP and Cohen GM. (1978) Metabolism of benzo(a)pyrene and its major metabolites to ethyl acetate-soluble and water-soluble metabolites by cultured rodent trachea. *Cancer Res* 38: 3066-3075.

Naslund I, Rubio CA and Auer GU. (1987) Nuclear-DNA Changes during Pathogenesis of Squamous Carcinoma of the Cervix in 3,4-Benzopyrene-Treated Mice. *Analytical and Quantitative Cytology and Histology* 9: 411-418.

Neal J and Rigdon RH. (1967) Gastric tumors in mice fed benzo(a)pyrene: a quantitative study. *Tex Rep Biol Med* 25: 553-557.

Neubert D and Tapken S. (1988) Transfer of Benzo(a)Pyrene into Mouse Embryos and Fetuses. *Archives of Toxicology* 62: 236-239.

Ng KM, Chu I, Bronaugh RL, *et al.* (1992) Percutaneous absorption and metabolism of pyrene, benzo[a]pyrene, and di(2-ethylhexyl) phthalate: comparison of in vitro and in vivo results in the hairless guinea pig. *Toxicol Appl Pharmacol* 115: 216-223.

NIEHS. (2000) Report on carcinogens : carcinogen profiles. In: Sciences. NIoEH (ed). Durham, N.C.: National Institute of Environmental Health Sciences, v.

Nielsen E, Østergaard G and Larsen JC. (2008) *Toxicological risk assessment of chemicals : a practical guide,* New York: Informa Healthcare.

NIOSH. (2012) Registry of toxic effects of chemical substances. In: Health NIfOSa (ed). Rockville, Md.: U. S. Dept. of Health, Education, and Welfare, Public Health Service, Center for Disease Control, National Institute for Occupational Safety and Health; Washington, for sale by the Supt. of Docs., v.

Norman J. (1992) One-Compartment Kinetics. *British Journal of Anaesthesia* 69: 387-396.

Notari RE. (1987) *Biopharmaceutics and clinical pharmacokinetics : an introduction,* New York: M. Dekker.

NPRI. (2012) 2012 NPRI National Database. National Pollutant Release Inventory.: Environment Canada.

NRC. (1987) *Drinking water and health,* Washington, D.C.: National Academy of Sciences.

NRC. (2006) Human biomonitoring for environmental chemicals. In: National Research Council (U.S.). Committee on Human Biomonitoring for Environmental Toxicants. and National Research Council (U.S.). Board on Environmental Studies and Toxicology. (eds). Washington, D.C.: National Academies Press,, xxi, 291 p.

OIT. (1986) *Principles of toxicokinetic studies,* Geneva: World Health Organization.

Osinski MA, Seifert TR, Cox BF, *et al.* (2002) An improved method of evaluation of drug-evoked changes in gastric emptying in mice. *J Pharmacol Toxicol Methods* 47: 115-120.

Paalzow LK and Teorell T. (1995) Torsten Teorell, the father of pharmacokinetics. *Upsala journal of medical sciences* 100: 41-46.

Pacheco JE. (2010) *Tarde o temprano : poemas, 1958-2009,* Barcelona: Tusquets Editores.

Payan JP, Lafontaine M, Simon P, *et al.* (2009) 3-Hydroxybenzo(a)pyrene as a biomarker of dermal exposure to benzo(a)pyrene. *Archives of Toxicology* 83: 873-883.

Paz O. (1969) *La centena : poemas, 1935-1968,* Barcelona: Barral Editores.

Perera FP, Hemminki K, Young TL, *et al.* (1988) Detection of polycyclic aromatic hydrocarbon-DNA adducts in white blood cells of foundry workers. *Cancer Res* 48: 2288-2291.

Perera FP, Tang DL, O'Neill JP, *et al.* (1993) HPRT and glycophorin A mutations in foundry workers: relationship to PAH exposure and to PAH-DNA adducts. *Carcinogenesis* 14: 969-973.

Pery AR, Brochot C, Desmots S, *et al.* (2011) Predicting in vivo gene expression in macrophages after exposure to benzo(a)pyrene based on in vitro assays and toxicokinetic/toxicodynamic models. *Toxicology Letters* 201: 8-14.

Peters SA. (2011) *Physiologically based pharmacokinetic (PBPK) modeling and simulations : principles, methods, and applications in the pharmaceutical industry,* Hoboken, N.J.: Wiley.

Poulin P and Krishnan K. (1995) An algorithm for predicting tissue: blood partition coefficients of organic chemicals from n-octanol: water partition coefficient data. *J Toxicol Environ Health* 46: 117-129.

Poulin P and Theil FP. (2000) A priori prediction of tissue:plasma partition coefficients of drugs to facilitate the use of physiologically-based pharmacokinetic models in drug discovery. *Journal of pharmaceutical sciences* 89: 16-35.

Pukkala EI. (1995) *Cancer risk by social class and occupation : a survey of 109,000 cancer cases among Finns of working age,* Basel ; New York: Karger.

Que Hee SS. (1993) *Biological monitoring : an introduction,* New York, NY: Van Nostrand Reinhold.

Ramesh A, Greenwood M, Inyang F, *et al.* (2001a) Toxicokinetics of inhaled benzo[a] pyrene: Plasma and lung bioavailability. *Inhalation Toxicology* 13: 533-555.

Ramesh A, Hood DB, Inyang F, *et al.* (2002) Comparative metabolism, bioavailability, and toxicokinetics of benzo[a]pyrene in rats after acute oral, inhalation, and intravenous administration. *Polycyclic Aromatic Compounds* 22: 969-980.

Ramesh A, Inyang F, Hood DB, *et al.* (2001b) Metabolism, bioavailability, and toxicokinetics of Benzo(alpha)pyrene in F-344 rats following oral administration. *Experimental and Toxicologic Pathology* 53: 275-290.

Reddy MB. (2005) *Physiologically based pharmacokinetic modeling : science and applications,* Hoboken, N.J.: Wiley-Interscience.

Redmond CK, Strobino BR and Cypess RH. (1976) Cancer Experience among Coke by-Product Workers. *Annals of the New York Academy of Sciences* 271: 102-115.

Rees ED, Mandelstam P, Lowry JQ, *et al.* (1971) A study of the mechanism of intestinal absorption of benzo(a)pyrene. *Biochim Biophys Acta* 225: 96-107.

Reisfeld B and Mayeno AN. (2012) *Computational toxicology,* New York: Humana Press ; Springer.

Rescigno A. (2010) Compartmental analysis and its manifold applications to pharmacokinetics. *AAPS J* 12: 61-72.

Rey-Salgueiro L, Garcia-Falcon MS, Martinez-Carballo E, *et al.* (2008) The use of manures for detection and quantification of polycyclic aromatic hydrocarbons and 3-hydroxybenzo[a] pyrene in animal husbandry. *Science of the Total Environment* 406: 279-286.

Robinson JR, Felton JS, Levitt RC, *et al.* (1975) Relationship between "aromatic hydrocarbon responsiveness" and the survival times in mice treated with various drugs and environmental compounds. *Molecular pharmacology* 11: 850-865.

Rodriguez LV, Dunsford HA, Steinberg M, *et al.* (1997) Carcinogenicity of benzo[alpha]pyrene and manufactured gas plant residues in infant mice. *Carcinogenesis* 18: 127-135.

Roth RA and Vinegar A. (1990) Action by the lungs on circulating xenobiotic agents, with a case study of physiologically based pharmacokinetic modeling of benzo(a)pyrene disposition. *Pharmacology & therapeutics* 48: 143-155.

Roth WL, Freeman RA and Wilson AG. (1993) A physiologically based model for gastrointestinal absorption and excretion of chemicals carried by lipids. *Risk Analysis* 13: 531-543.

Rowland M and Tozer TN. (2011) *Clinical pharmacokinetics and pharmacodynamics : concepts and applications,* Philadelphia, Pa. ; London: Lippincott Williams & Wilkins.

Rubinow SI. (2002) *Introduction to mathematical biology,* Mineola, New York: Dover Publications.

Rutherford E. (1962) *The collected papers of Lord Rutherford of Nelson, O.M.F.R.S. : Published under the scientific direction of James Chadwick,* London: Allen and Unwin.

San Jose R, Perez JL, Callen MS, *et al.* (2013) BaP (PAH) air quality modelling exercise over Zaragoza (Spain) using an adapted version of WRF-CMAQ model. *Environmental Pollution* 183: 151-158.

Sanders CL, Skinner C and Gelman RA. (1984) Percutaneous-Absorption of [Benzo[a]Pyrene-7.10-C-14 and [7,12-C-14]Dimethylbenz[a]Anthracene in Mice. *Environmental Research* 33: 353-360.

Santos FJ and Galceran MT. (2002) The application of gas chromatography to environmental analysis. *Trac-Trends in Analytical Chemistry* 21: 672-685.

SCF. (2002) Polycyclic Aromatic Hydrocarbons – Occurrence in foods, dietary exposure and health effects. In: Food SCo (ed) *HEALTH and CONSUMER PROTECTION DIRECTORATE-GENERAL.* Directorate C - Scientific Opinions: EUROPEAN COMMISSION.

Schlede E, Kuntzman R, Haber S, *et al.* (1970) Effect of enzyme induction on the metabolism and tissue distribution of benzo(alpha)pyrene. *Cancer Res* 30: 2893-2897.

Schuler LJ, Wheeler M, Bailer AJ, *et al.* (2003) Toxicokinetics of sediment-sorbed benzo[a]pyrene and hexachlorobiphenyl using the freshwater invertebrates Hyalella azteca, Chironomus tentans, and Lumbriculus variegatus. *Environmental Toxicology and Chemistry* 22: 439-449.

Schulte A, Ernst H, Peters L, *et al.* (1994) Induction of squamous cell carcinomas in the mouse lung after long-term inhalation of polycyclic aromatic hydrocarbon-rich exhausts. *Experimental and toxicologic pathology : official journal of the Gesellschaft fur Toxikologische Pathologie* 45: 415-421.

Sellgren K. (2001) Aromatic hydrocarbons, diamonds, and fullerenes in interstellar space: puzzles to be solved by laboratory and theoretical astrochemistry. *Spectrochimica Acta Part a-Molecular and Biomolecular Spectroscopy* 57: 627-642.

Seubert JM and Kennedy CJ. (2000) Benzo[a]pyrene toxicokinetics in rainbow trout (Oncorhynchus mykiss) acclimated to different salinities. *Archives of Environmental Contamination and Toxicology* 38: 342-349.

Shimmins J, Allison AC, Smith DA, *et al.* (1967) A two-compartment kinetic model: rigorous and approximate solutions. *Calcified tissue research* 1: 137-143.

Shu HP and Bymun EN. (1983) Systemic excretion of benzo(a)pyrene in the control and microsomally induced rat: the influence of plasma lipoproteins and albumin as carrier molecules. *Cancer Res* 43: 485-490.

Shubik P, Pietra G and Della Porta G. (1960) Studies of skin carcinogenesis in the Syrian golden hamster. *Cancer Res* 20: 100-105.

Shum S, Jensen NM and Nebert DW. (1979) The murine Ah locus: in utero toxicity and teratogenesis associated with genetic differences in benzo[a]pyrene metabolism. *Teratology* 20: 365-376.

Sims P, Grover PL, Swaisland A, *et al.* (1974) Metabolic activation of benzo(a)pyrene proceeds by a diol-epoxide. *Nature* 252: 326-328.

Sparnins VL, Mott AW, Barany G, *et al.* (1986) Effects of allyl methyl trisulfide on glutathione S-transferase activity and BP-induced neoplasia in the mouse. *Nutr Cancer* 8: 211-215.

Spengler JD, MacIntosh DL, WHO Task Group on Human Exposure Assessment, *et al.* (2000) *Human exposure assessment,* Geneva: World Health Organization.

Spinelli JJ, Demers PA, Le ND, *et al.* (2006) Cancer risk in aluminum reduction plant workers (Canada). *Cancer Causes & Control* 17: 939-948.

Stern AH. (1997) Estimation of the interindividual variability in the one-compartment pharmacokinetic model for methylmercury: Implications for the derivation of a reference dose. *Regulatory Toxicology and Pharmacology* 25: 277-288.

Storm JE, Collier SW, Stewart RF, *et al.* (1990) Metabolism of xenobiotics during percutaneous penetration: role of absorption rate and cutaneous enzyme activity. *Fundam Appl Toxicol* 15: 132-141.

Sun JD, Wolff RK and Kanapilly GM. (1982) Deposition, Retention, and Biological Fate of Inhaled Benzo(a)Pyrene Adsorbed onto Ultrafine Particles and as a Pure Aerosol. *Toxicology and Applied Pharmacology* 65: 231-244.

Szczeklik A, Szczeklik J, Galuszka Z, *et al.* (1994) Humoral immunosuppression in men exposed to polycyclic aromatic hydrocarbons and related carcinogens in polluted environments. *Environ Health Perspect* 102: 302-304.

Tardiff RG and Goldstein BD. (1991) *Methods for assessing exposure of human and non-human biota,* Chichester, West Sussex, England ;

Toronto: J. Wiley.

Thompson MD and Beard DA. (2011) Development of appropriate equations for physiologically based pharmacokinetic modeling of permeability-limited and flow-limited transport. *Journal of Pharmacokinetics and Pharmacodynamics* 38: 405-421.

Thyssen J, Althoff J, Kimmerle G, *et al.* (1981) Inhalation studies with benzo[a]pyrene in Syrian golden hamsters. *Journal of the National Cancer Institute* 66: 575-577.

Timbrell JA. (2002) *Introduction to toxicology,* London ; New York: Taylor & Francis.

Tornqvist S, Norell S, Ahlbom A, *et al.* (1986) Cancer in the Electric-Power Industry. *British Journal of Industrial Medicine* 43: 212-213.

Toth B. (1980) Tumorigenesis by Benzo(a)Pyrene Administered Intracolonically. *Oncology* 37: 77-82.

Tozer TN and Rowland M. (2006) *Introduction to pharmacokinetics and pharmacodynamics : the quantitative basis of drug therapy,* Philadelphia: Lippincott Williams & Wilkins.

Van Hattum B and Montanes JFC. (1999) Toxicokinetics and bioconcentration of polycyclic aromatic hydrocarbons in freshwater isopods. *Environmental Science & Technology* 33: 2409-2417.

Verschueren K. (2001) Handbook of environmental data on organic chemicals. 4th ed. New York: Wiley,, 2 v.

Wagner JG. (1981) History of pharmacokinetics. *Pharmacology & therapeutics* 12: 537-562.

Weigt S, Huebler N, Strecker R, *et al.* (2011) Zebrafish (Danio rerio) embryos as a model for testing proteratogens. *Toxicology* 281: 25-36.

Wells JD. (2012) Effective theories in physics from planetary orbits to elementary particle masses. *SpringerBriefs in physics,.* Berlin ; New York: Springer,, 1 online resource.

Weyand EH and Bevan DR. (1986) Benzo(a)pyrene disposition and metabolism in rats following intratracheal instillation. *Cancer Res* 46: 5655-5661.

White RK. (1988) Pharmacokinetics in Risk Assessment - Drinking-Water and Health, Vol 8 - Natl-Res-Council-Commiss-Life-Sci-Board-Environm-Studies-and-Toxicol-Safe-Drinking-Water-Comm-Subcomm-Pharmacokinet-Risk-Assessment. *Risk Analysis* 8: 305-305.

WHO. (1998) Selected non-heterocyclic polycyclic aromatic hydrocarbons. In: WHO Task Group on Environmental Health Criteria for Selected Non-Heterocyclic Polycyclic Aromatic Hydrocarbons (ed) *Environmental health criteria,.* Geneva: World Health Organization,, 1 online resource.

WHO. (2000) *Air quality guidelines for Europe,* Copenhagen: World Health Organization, Regional Office for Europe.

Wiersma DA and Roth RA. (1983) Total-Body Clearance of Circulating Benzo(a)Pyrene in Conscious Rats - Effect of Pretreatment with 3-Methylcholanthrene and the Role of Liver and Lung. *Journal of Pharmacology and Experimental Therapeutics* 226: 661-667.

Withey JR, Shedden J, Law FC, *et al.* (1993) Distribution of benzo[a]pyrene in pregnant rats following inhalation exposure and a comparison with similar data obtained with pyrene. *Journal of applied toxicology : JAT* 13: 193-202.

Wojdani A and Alfred LJ. (1984) Alterations in cell-mediated immune functions induced in mouse splenic lymphocytes by polycyclic aromatic hydrocarbons. *Cancer Res* 44: 942-945.

Wolff RK, Griffith WC, Henderson RF, *et al.* (1989) Effects of repeated inhalation exposures to 1-nitropyrene, benzo[a]pyrene, Ga2O3 particles, and SO2 alone and in combinations on particle clearance, bronchoalveolar lavage fluid composition, and histopathology. *J Toxicol Environ Health* 27: 123-138.

Wood AW, Levin W, Lu AYH, *et al.* (1976) Metabolism of Benzo[a]Pyrene and Benzo[a]Pyrene Derivatives to Mutagenic Products by Highly Purified Hepatic Microsomal-Enzymes. *Journal of Biological Chemistry* 251: 4882-4890.

Xu ZY, Brown LM, Pan GW, *et al.* (1996) Cancer risks among iron and steel workers in Anshan, China .2. Case-control studies of lung and stomach cancer. *American Journal of Industrial Medicine* 30: 7-15.

Yamagiwa K and Ichikawa K. (1977) Classics in Oncology - Yamagiwa,K Experimental-Study of Pathogenesis of Carcinoma. *Ca-a Cancer Journal for Clinicians* 27: 172-181.

Yamamoto S, Natsumeda S, Hara K, *et al.* (2014) Applicability of Concentrations Obtained by Working Environment Measurement to Assessment ofPersonal Exposure Concentrations of Chemicals. *J Occup Health.*

Yang HY, Majesky MW, Namkung MJ, *et al.* (1986a) Phase II biotransformation of carcinogens/atherogens in cultured aortic tissues and cells. II. Glucuronidation of 3-hydroxy-benzo(a)pyrene. *Drug metabolism and disposition: the biological fate of chemicals* 14: 293-298.

Yang HY, Namkung MJ, Nelson WL, *et al.* (1986b) Phase II biotransformation of carcinogens/atherogens in cultured aortic tissues and cells. I. Sulfation of 3-hydroxy-benzo(a)pyrene. *Drug metabolism and disposition: the biological fate of chemicals* 14: 287-292.

Yang SK, Selkirk JK, Plotkin EV, *et al.* (1975) Kinetic analysis of the metabolism of benzo(a)pyrene to phenols, dihydrodiols, and quinones by high-pressure chromatography compared to analysis by aryl hydrocarbon hydroxylase assay, and the effect of enzyme induction. *Cancer Res* 35: 3642-3650.

Yang SK and Silverman BD. (1988) *Polycyclic aromatic hydrocarbon carcinogenesis : structure-activity relationships,* Boca Raton, Fla.: CRC Press.

Yoon M, Efremenko A, Blaauboer BJ, et al. (2014) Evaluation of simple in vitro to in vivo extrapolation approaches for environmental compounds. Toxicology in Vitro 28: 164-170.

Zeilmaker MJ, Eijkeren JC and Olling M. (1997) A PBPK-model for B(a)P in the rat relating dose and liver DNA-adduct level. *RIVM Report 658603 008.* Bilhoven, The Netherlands.

Printed by Books on Demand GmbH, Norderstedt / Germany